Edible & Medicinal Plants of the Northwest

Alaska, Western Canada & the
Northwestern United States

J. Duane Sept

Calypso Publishing

Front cover photos by J. Duane Sept: Clockwise from top left; blue clematis, heart-
leaved arnica, red-osier dogwood, Canada buffaloberry, western skunk cabbage,
western yew, Nootka rose, common bearberry, red clover (center), fireweed (back-
ground)
Back cover photos by J. Duane Sept: broad-leaved stonecrop, red osier dogwood

Calypso Publishing, P.O. Box 1141, Sechelt, BC, Canada, V0N 3A0
www.calypso-publishing.com

Duane Sept Photography: www.septphoto.com

Printed in China.

Library and Archives Canada Cataloguing in Publication

Sept, J. Duane, 1950-, author
        Edible and medicinal plans of the Northwest : Alaska, Western
Canada and the Northwestern United States / J. Duane Sept.

ISBN 978-0-9739819-8-8 (pbk.)
        1. Wild plants, Edible--Canada, Western--Identification. 2. Wild
plants, Edible--Alaska--Identification. 3. Wild plants, Edible--West (U.S.)--
Identification. 4. Medicinal plants--Canada, Western--Identification.
5. Medicinal plants--Alaska--Identification. 6. Medicinal plants--West (U.S.)--
Identification. I. Title.

QK98.5.S46 2014   581.6'3209711   C2013-907792-8

CAUTION!
The purpose of this guide is not to encourage the reader to eat a plant or use
it medicinally, but rather to learn about and appreciate the history of plant
use by Native peoples and others. Many species of plants in the Northwest
are poisonous—and some are deadly. Other species, including edible ones,
may cause allergic reactions or other unpredictable physical responses. All
information in this book is accurate to the best of the author's knowledge.
The author, publisher, distributor and retailer assume no liability for the
actions of the reader. Many of the plants listed here are toxic and must not
be ingested. If you wish to use a plant medicinally, always seek advice from
your physician or a qualified herbal specialist first. **Be cautious.**

Contents

Introduction

The use of plants for medicinal purposes is known to be centuries old. Some plants, like common yarrow, were used in the Middle Ages in Europe to stop internal bleeding. Native peoples of the Northwest also have a long history of using the plants growing around

them to aid them in their day-to-day life. Some of these remedies have been verified scientifically to be of tremendous medicinal value in the treatment of diseases. For example, the western yew *Taxus brevifolia* was found to have a chemical in its bark, taxol, a compound effective in the treatment of ovarian and breast cancer. Taxol (also a trade name) has even been approved by the US Food and Drug Administration and elsewhere around the globe as a cancer treatment.

Edible plants too have had an important history. Some species are well known for their edibility while others are lesser known. The common edible berries and other fruits of the Northwest are not included in this guide, as information on these species is readily available in many other references on wild fruits and berries.

Where to Collect Plants

Depending upon your location, there may be many sites near you, including government forests or unprotected local areas. Some species may even occur in your own backyard! Keep in mind, however, that you should never collect plants that may have been sprayed with herbicides. These should be considered toxic. Also, never collect plants from areas where the water may be polluted. Prudence is always the best policy.

How to Use This Guide

This book gives information on and helps identify edible and medicinal plants that grow throughout the Northwest.

Species
A common and a scientific name are listed for each species. Every living organism has a unique scientific name consisting of two parts: the genus (a grouping of species with common characteristics) and the species. Occasionally names change as new scientific information is discovered. The most current or appropriate name is included in this book.

Common names are those used in everyday conversation by people who live in an area where the species occur, so many plants have several common names. The most widely accepted common name appears at the top of each entry with the species' scientific name.

Other Names
Other common and scientific names for the species are listed here.

Family
A plant family is a grouping of one or more genera with similar overall characteristics. All species of *Pinus*, for instance, belong to the pine family Pinaceae, which includes many genera.

Description
To identify a species, use the photograph and the written description together. Data on the flowers, fruits, leaves and other features accompany a description of the plant. Together these features will help you identify plant species.

Size
The maximum size of each species is shown.

Habitat
Habitat is the type of surroundings in which a species normally grows. Many plants are found in more than one habitat; some have specific requirements for moisture or other conditions.

Range
Range is the physical area in which the species is known to grow.

Edible Uses
The edible parts of the plant are identified in this section.

Medicinal Uses
The medicinal parts of the plant are identified in this section.

Precautions
Any known concerns about the plant, such as its poisonous parts, are included here. Sometimes health concerns and side effects are also noted. In general, however, it is important to note that large quantities of any edible or medicinal plant are not recommended, as they may be toxic. Pregnant women should be particularly cautious.

Notes
Notes accompanying each species give special information, such as interesting features of the plant.

Similar Species
Species that are similar in appearance are identified, along with notes and additional information.

Preparation

Collecting plants

Where you collect your plants is always an important consideration. Collect away from roads or other areas where plants could be subject to dust, pesticides or vehicle exhaust. Be sure that the surrounding area is not the site of a mine or other activity that might compromise the plant's air, water or soil. Plants concentrate toxins. Pick only what you will use. Overharvesting pressure on some species can quickly reduce their numbers. Always consider the environment!

Drying plants

Cut the stems just below the green leaves and bundle the plants together with rubber bands. Hang them outdoors, out of direct sunlight, in a sheltered area with good ventilation. Leave them there until all parts are totally dry. In areas of high humidity, outdoor drying rarely works. There it is best to dry the plants indoors, using a dehydrator. Do not use an oven or microwave.

Herbal Preparations

Several preparations have been developed over the centuries to make the most of the properties of various plants. The following techniques are the most common.

Decoction
A decoction is a water solution of plant extracts that uses both fine and coarse plant components (roots, bark or chips) by boiling them. To prepare a decoction, grind about 2 oz (50 g) of the plant. Place it in a pot made from a non-reactive metal (such as stainless steel, not aluminum). Add 1 qt (1 L) of cold water, bring the mixture to a boil and simmer for 20 minutes. Once cool, strain the decoction and place in a container. Keep it in the refrigerator and use it within 24 hours.

Infusion
An infusion is a water solution that uses only delicate plant components (including the buds, leaves or flowers), which are steeped in hot or boiling water for a short time. To make an infusion, boil 32 parts of water by weight and remove it from the heat. Then add 1 part by weight of the dried plant and let stand for up to 1 hour. Strain the infusion into a container. Then add water (to replace the water that evaporated) to reach the 32 parts originally measured. Keep the infused liquid in the refrigerator and use within 24 hours. Cold infusions can also be made. Suspend 1 part, by weight, of the dried plant in a cloth or paper towel into 32 parts water and let steep for at least 6 hours to allow the soluble solids to dissolve.

Tincture
A tincture is a fluid plant extract made with alcohol or glycerin, which allows it to last a long time. A typical tincture is prepared by steeping 1 part fresh herb in 2 parts alcohol or glycerin for 10 to 14 days. It is important to use pure grain alcohol (ethanol), 190 proof (95% alcohol). This is available in most provinces in Canada (not BC) and most states in the US (not California). Lower-proof alcohol does not produce acceptable results. It is also possible to make a dry tincture using dry powder of the herb rather than the fresh herb. This tincture must be mixed twice each day for 14 days. The weight of herb to use varies with each preparation. Consult the medicinal resources listed in the Bibliography (p. 88) for details.

Poultice
A poultice is normally a moist herbal mass, warm to hot, that is laid on some part of the body. A cloth is often used to keep the mass moist and contained. Some kinds of poultice can be left on an affected area for extended periods of time if the skin is not irritated by them.

EDIBLE & MEDICINAL PLANTS

Western Redcedar *Thuja plicata*

Other Names Also known as arbor vitae western red cedar, Pacific redcedar, giant cedar, shinglewood.

Cypress Family (Cupressaceae)

Description Overall: Coniferous tree.

Cones: Pollen cones reddish and very small, to 0.2" (4 mm) long. Seed cones greenish, brown when mature at 2 years. Cones are egg-shaped with 8–12 scales in woody, irregular clusters that drop during the winter months. **Leaves:** Evergreen; yellowish green, scale-like, flattened to 0.25" (6 mm) long in 4 rows on small twigs attached to vigorously growing branches. **Other:** Tree stems from a relatively shallow but widespread root system.

Size Normally to 131' (40 m) high, occasionally to 197' (60 m).

Habitat Primarily in moist to wet soils, in shaded forests; low to mid-elevation.

Range Alaska to northwest California, east to the Rocky Mountains.

Medicinal Uses Western redcedar is well known for its antifungal and antibacterial properties, although it has not been used regularly for medicinal purposes. A fresh tincture

(glycerin) is sometimes used to treat athlete's foot, ringworm and similar afflictions by being applied two or three times per day for a full week. An infusion of bark and twigs has been given to patients with kidney problems. Chronic bladder and urethral irritability has also been treated with a cold infusion of western redcedar twice a day. The buds of the cedar have been chewed for toothache. The oil of western redcedar has been applied externally to warts, fungus infections and piles.

Precautions Western redcedar should not be used for extended periods by those with kidney weakness. It should not be taken internally during pregnancy.

The oil in the leaves is toxic! It can cause low blood pressure and convulsions, and has even caused death.

Notes Western redcedar is more famous for its value in construction than in medicine. The wood is world-renowned for its properties both in modern construction and in historical building by Native peoples.

Seed cones.

White-bark Pine *Pinus albicaulis*

Other Names Also known as scrub pine, white pine, whitestem, alpine whitebark, pitch pine, scrub pine, creeping pine.

Pine Family (Pinaceae)

Description Overall: Coniferous tree. **Cones:** Pollen cones scarlet; to 0.6" (15 mm) long. Seed cones purplish, maturing to brown; to 3.3" (8 cm) long; permanently closed—do not open on drying; grow at right angles to the branch; often very pitchy; seed crops produced at irregular intervals; mature August to September. **Leaves:** Evergreen; needles in bundles of 5; to 3.5" (9 cm) long. **Other:** Plants grow from a deep and spreading root system.

Size To 50' (15 m) high.

Habitat Exposed slopes and ridges, subalpine to timberline.

Range Central BC to central California, east to Idaho and Wyoming.

Edible Uses The edible seeds in the cones of the white-bark pine are edible and traditionally were a favorite food for Native peoples. Cones were gathered in September, when closed tightly, and spread out on the ground so that the sun would dry them out and open them. The seeds were extracted by knocking them against a hard object. They were sometimes eaten raw, but normally roasted to soften the bitter taste. They were sometimes cooked, mixed with berries and stored for winter.

Medicinal Uses The Navajo people used pine-needle tea to treat fever and cough. This tea is also believed to initiate urination and to aid in bringing up phlegm and mucus. The resin of pines has long been an ingredient in cough syrups, as well as ointments for burns and skin infections. The tender young needles are an excellent source of vitamin C.

Precautions If many seeds are eaten raw, they can cause constipation, so it is best to roast them first. Evergreen tea can be toxic if consumed in large amounts. Extended use can irritate the kidneys. Pregnant women should not drink this tea at all.

Notes All pine trees have a life-sustaining edible inner bark and the all-important vitamin C. The inner bark is also a favorite of bears.

Seed cones.

Lodgepole Pine *Pinus contorta*

Other Names Also known as black pine, scrub pine, tamarack pine, mountain pine.

Pine Family (Pinaceae)

Description Overall: Coniferous tree. **Cones:** Pollen cones reddish green to yellow, small; in clusters at branch tips in spring. Seed cones brown; to 2" (5 cm) long; hard, armed with prickles; usually paired, curved, pointing away from branch; usually stay closed and on the tree for 10–20 years; good crops every 1–3 years. **Leaves:** Evergreen; needles in bundles of 2; to 2.8" (7 cm) long. **Other:** Tree normally grows from shallow root, but taproot and vertical sinkers are often present on well-drained sites.

Size Normally to 82' (25 m) high; occasionally to 131' (40 m).

Habitat Sites ranging from rock outcrops to areas with deep, rich organic soils; sea level to subalpine.

Range Alaska to Oregon, east to Alberta.

Edible Uses The inner bark is sweet and succulent, and was considered an important food by the Interior aboriginal peoples. It was normally eaten fresh, but the Secwepemc also dried it for use later in the year. The inner bark is best harvested in late May and June when the sap is running.

Medicinal Uses The resin of the lodgepole, as with all pines, was used for treating sore muscles and symptoms of rheumatoid arthritis, as well as relieving the soreness and inflammation of wounds and cuts. It was also heated and smeared on the chest for pneumonia.

Native peoples used pine oil and tar as an antiseptic, disinfectant and insecticide, as an agent for parasite removal, and for many other purposes. The pitch of the lodgepole pine, like that of other conifers, is often directly applied to small cuts and abrasions to promote healing and reduce scarring. A root decoction of lodgepole pine has been used to clean wounds and aid healing. A poultice of softened inner bark has also been

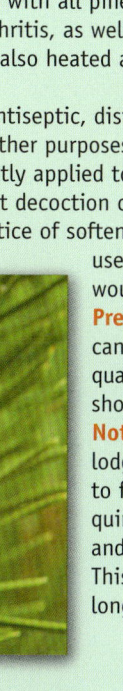

used to help heal more serious wounds.

Precautions Tea from this pine can be toxic if taken in large quantities. Pregnant women should not drink this tea.

Notes The reproduction of the lodgepole pine is closely linked to fire ecology. The cones require a hot, intense fire to open and release the seeds inside. This species is known to live as long as 300 years.

Seed cone.

Western Larch *Larix occidentalis*

Other Names Also known as western tamarack, tamarack, larch, mountain larch, hackmatack.

Pine Family (Pinaceae)

Description Overall: Coniferous tree that sheds its needles annually. **Cones:** Pollen cones yellow; to 0.4" (1 cm) long. Seed cones reddish when young, brown when mature; to 1.6" (4 cm) long; erect; long bract tips extend beyond scales. Seed production begins as early as 15 years, and may continue for 200 years or more. **Leaves:** Deciduous; needles in tufts of 15–30; 3-sided; to 2" (5 cm) long; shed for winter; needles are triangular in cross-section; twigs are slightly hairy. **Other:** Tree grows from a deep, extensive root system.

Size Normally to 150' (46 m) high; occasionally to 230' (70 m).

Habitat Moist to dry sites, often gravelly or sandy; foothills to montane.

Range BC south to Oregon; east to Alberta and Montana.

Edible Uses Native peoples ate the sweet inner bark of this tree in the springtime. The western larch produces a gum that hardens on exposure to air. This gum is naturally sweet-tasting and it could be found and chewed all year round. It was also used as a leavener, similar to baking powder. The trees have also been tapped for their sap, which was condensed to produce a syrup.

Medicinal Uses The bark of the western larch contains high levels of larch arabinogalactan, believed to boost the immune system and help protect individuals against the common cold, influenza, hepatitis C, HIV/AIDS and other viral and bacterial ailments.

Notes The larches are well known for turning a golden color in autumn, just before they lose their needles.

Similar Species Subalpine Larch *Larix lyallii* is a similar species that is found at timberline.

Subalpine Fir *Abies lasiocarpa*

Other Names Also known as alpine fir, Rocky Mountain fir.

Pine Family (Pinaceae)

Description **Overall:** Coniferous tree. **Cones:** Pollen cones are bluish and small. Seed cones are light to dark purple and grow to 4" (10 cm) long; erect with bracts dropping while still on the tree. Good cone crops occur about once every 3 years. **Leaves:** Evergreen; needles flat, to 1.5" (4 cm) long; needles do not lie flat on branch; white lines of stomata (tiny pores used for gas exchange) on upper and lower surfaces. **Other:** Trees arise from deep or shallow roots depending upon soil drainage.

Size Normally to 75' (23 m) high; occasionally to 150' (45 m).

Habitat Mountain slopes; sea level to subalpine.

Range Southeast Alaska to New Mexico.

Edible Uses The Blackfoot people gathered the cones left behind by chipmunks and squirrels. They pulverized them into a powder, mixed it with back fat or marrow, cooled it and served it as a delicacy at social events.

Medicinal Uses Subalpine fir was an important medicinal plant for Native peoples. Lewis and Clark also kept it in their medicine kit and used it (both raw and as an infusion) for asthma, cough and colds. It was also used in salves to treat cuts, wounds, ulcers, sores, bleeding gums, skin infections and other maladies.

The needles were ground very fine and used in the treatment of open sores. The resin or needles of subalpine fir were added to poultices used to treat fever and chest colds. These ailments were also treated with resin or needle tea. A needle infusion was given to people who were coughing up blood or showing other signs of tuberculosis.

Precautions Some individuals' skin has an allergic reaction to fir resin. Moderation is the key for this and all evergreen teas. Do not eat the needles.

Notes The resin of subalpine fir had many traditional uses including as a remedy for bad breath. The needles were sometimes placed inside small bags and hung around a horse's neck as a perfume.

Seed cone.

Western Yew *Taxus brevifolia*

Other Names Also known as Pacific yew, mountain mahogany, yew.
Yew Family (Taxaceae)
Description Overall: Evergreen tree.
Flowers: Male: cylindrical cones with yellowish pollen sacs. Female: minute green flowers to 0.1" (3 mm) long; June. **Fruits:** Salmon to scarlet, cup-shaped, covers a large brown seed; to 0.4" (9 mm) across; August to October. **Leaves:** Flat needles; to 1.5" (3.5 cm) long. **Other:** Plants arise from a deep, wide-spreading root system.
Size To 80' (24 m) high.
Habitat Moist mature forest, common along the coast; low to mid-elevation.
Range Southern Alaska to Idaho, south to California.
Medicinal Uses The western yew has recently been found to contain a potent anti-cancer agent, called Taxol (also a trade name), known to be effective in the treatment of ovarian and breast cancer. Its use is approved by the US Food and Drug Administration and by regulatory agencies elsewhere.

First Nations groups have used this bark as a medicine for a very long time. It was one of four ingredients in a medicine that was taken for ulcers, liver ailments and other internal conditions. Several Native tribes made a sedative from the needles and twigs. Like other plants, western yew should only be administered by people who are experienced in preparing it. Always check with a qualified herbal specialist or physician before trying any kind of plant medicine. **Caution is advised!**

Precautions All parts of this plant, including the fruit, are considered **toxic**. The poison is an alkaloid that alters the rhythm of the heartbeat and can cause sudden heart failure.

Notes Taxol is present throughout the entire tree, but it is most concentrated in the bark. There the concentration is still small—14 kg of the bark yields a mere 1 gram of Taxol, and one treatment requires 2 grams (which would require approximately 10 good-sized trees). The demand for its bark has therefore resulted in a large-scale reduction in western yew. A synthetic compound is currently in development.

Paper Birch *Betula papyrifera*

Other Names Also known as canoe birch, silver birch, white birch.

Birch Family (Betulaceae)

Description **Overall:** Deciduous tree. **Flowers:** Male and female catkins present on the same tree, female catkins shorter and thinner than male catkins; to 1.6" (4 cm) long. **Fruits:** Mature seed catkins with nutlets inside; catkins to 1.5" (3.8 cm) long; break apart when ripe. Leaves: Simple, alternate, deciduous; pale green, turning yellow in autumn; ovate with a pointed tip; margins double-toothed; to 4" (10 cm) long. **Other:** This tree grows from a very shallow root system.

Size Normally to 100' (30 m) high.

Habitat A variety of forests, usually on moist, well-drained sites and around bogs and other wetlands; also a pioneer species in burned areas; low to mid-elevation.

Range Northwest Alaska to Oregon; east to Newfoundland and south to New York.

Edible Uses A syrup can be made from birch sap, but 20–26 gallons (80–100 L) of sap are required to make one quart (1 L) of syrup. The sap can also be consumed as a health food in its natural state, and can be used to make birch beer. The inner bark has been eaten in early spring as both a special treat and a starvation food.

Medicinal Uses The inner bark was boiled and made into a poultice in treating burns and other wounds. Paper birch leaves were chewed and laid on top of wasp stings to extract the poison. Gonorrhea has been treated with the buds or wood of the paper birch. Buds were mixed with lard to make an ointment for treating skin sores and infections. Another ointment, for sores and rashes, was made from the inner bark mixed with pitch and grease.

Notes The bark of the paper birch was an important material, especially for interior Native peoples, who used it in making canoes and many other products.

Seed catkin.

Red alder.

Alder *Alnus* spp.

Birch Family (Betulaceae)

Description **Overall:** Deciduous tree or shrub. **Flowers:** Pollen catkins in small clusters at shoot tips; male yellowish to 6" (15 cm) long; female reddish to 0.5" (1.2 cm) long and very narrow. **Fruits:** Nutlet is narrowly winged, inside cone. **Leaves:** Deciduous; dark green above; ovate to elliptical; to 5" (12.5 cm) long; teeth along edge. **Other:** Plants arise from shallow roots that are wide-spreading like those of legumes, often with swellings or nodules.

Size To 100' (30 m) high.

Habitat Moist, open areas such as stream banks, valley bottoms and floodplains.

Range Alaska to California.

Edible Uses The inner bark of red alder was eaten in the spring by several Northwest Coastal aboriginal groups. The young buds have also been noted as edible.

Medicinal Uses Alder is known to have strong antibiotic properties and is considered an astringent. A bark decoction was prepared, aged and used in the treatment of constipation, jaundice and diarrhea. Alder bark tea was used to treat hemorrhoids. A bark solution was used for respiratory ailments, as a tonic and as a wash for skin infections and wounds.

The root of the alder was boiled and made into a tea by Native peoples as an astringent to aid individuals who had blood in their stools. Itchiness caused by rashes has been treated with the juice of freshly scraped alder bark. Sexually transmitted disease in men was treated with a tea made from the boiled female catkins of the green alder.

Leaf decoctions have been used on burns and inflamed wounds. A fresh infusion is reported to be quite effective in treating poison ivy inflammation. A poultice of the inner bark has been used to reduce swellings and to treat wounds and skin ulcers.

Notes Several species of alder are present in the Northwest, including mountain alder *Alnus incana tenuifolia*, red alder *Alnus rubra*, green alder *Alnus viridis crispa*, Sitka alder *Alnus viridis sinuata* and white alder *Alnus rhombifolia*.

Alder is an excellent dye plant that gives us wonderful red, gold, olive green, brown and black dyes. It is also important ecologically as a nitrogen-fixing plant in newly colonized sites.

Balsam Poplar *Populus balsamifera*

Other Names Also known as balm buds, balm of Gilead, balsam, black cottonwood, black poplar, California poplar, Carolina poplar, cottonwood, hackmatack poplar, poplar balsam, rough barked poplar, tacamahac, tacamahac poplar, tackamahac, western balsam poplar; formerly classified as *P. trichocarpa*.
Willow Family (Salicaceae)
Description Overall: Deciduous tree. **Flowers:** Very small, in long, slender, loosely hanging catkins; sexes on separate trees appearing before leaves unfold; male catkins to 1.2" (3 cm) long, 20–30 stamens; female catkins to 4" (10 cm) long, 2 stigmas. **Fruits:** Green capsules, egg-shaped, in hanging catkins; to 0.3" (8 mm) long; splits into 2–3 parts when ripe; seed very small with a tuft of cottony hairs, dispersed in large masses by the wind. **Leaves:** Simple, alternate, deciduous; shiny, dark green above, silvery below, ovate, pointed at tip, rounded to heart-shaped at base; leaf stalk round, to 6" (15 cm) long. **Other:** Plants arise from a multi-layered root system.

Size To 82' (25 m) high.

Habitat River valleys, stream banks, floodplains and similar low-lying sites.

Range Alaska to California and New Mexico.

Edible Uses The cambium has been scraped off and eaten by hunters, and it has also been used as a starvation food in the summer months.

Medicinal Uses The sap has been drunk to treat diabetes and high blood pressure. The bark and sap have been boiled together to make a tea to treat symptoms of asthma in children. Balm of Gilead, made from the bark or buds of balsam poplar and several other poplar species, has been found effective in treating bronchitis, cleansing the blood and eliminating the cause of scurvy. It has also been used as a bath to treat skin diseases such as eczema and psoriasis. Buds were collected in late winter and early spring in order to make Balm of Gilead. It was common to place a bud of balsam poplar into a nostril to stop a nosebleed. A drink made by boiling the buds with aspen poplar branch bark for 30 minutes has been used to treat diabetes. The leaves have been applied to sores to draw out infection. A salve was made by crushing the sticky leaf buds in lard, then spread on sunburn, scalds, scratches and wounds. The leaves have been used to treat bruises, sores, boils and aching muscles. Compresses made of mashed leaves were applied to treat headaches. Poultices of raw or boiled roots were used to treat sprains and bruises.

Precautions There are reports indicating that the bark tea may be slightly toxic. Those with sensitive skin may find bud resin an irritant.

Notes Balsam poplar is a common, easily identified tree that is used for a wide variety of ailments.

Similar Species Trembling Aspen *Populus tremuloides* (see p. 20)

Trembling Aspen *Populus tremuloides*

Other Names Also known as aspen, quaking aspen, white poplar, golden aspen, mountain aspen, popple, poplar, trembling poplar.
Willow Family (Salicaceae)
Description Overall: Deciduous tree.
Flowers: Male and female on separate trees, appearing before leaves; catkins to 4" (10 cm) long. Male flowers with 6–12 stamens, bracts fringed with long hairs. Female flowers with 2 stigmas, bracts fringed with long hairs. **Fruits:** Cone-shaped capsules to 0.3" (7 mm) long; catkins to 4" (10 cm) long; many very small cotton-tipped seeds in each capsule. **Leaves:** Simple, alternate, deciduous; fresh green above, paler beneath; round to heart-shaped, abruptly tipped, finely round-toothed, stalks long, slender, flattened; to 3" (7.5 cm) long. **Other:** Plants arise from spreading roots.
Size Normally to 60' (18 m) high; occasionally to 100' (30 m).

Habitat Dry to moist sites, often in pure stands; low elevation to subalpine.
Range Alaska to California and New Mexico.
Edible Uses The inner bark was eaten as an emergency food by Native peoples.
Medicinal Uses The inner bark of trembling aspen contains a compound, salicin, which is very similar to salicylic acid, and this was used much as we use aspirin today. Those who try the inner bark of trembling aspen will find that the taste leaves a lot to be desired. Aboriginal groups prepared a tea from the bark and used it to treat a wide range of ailments including skin problems, fevers, jaundice and diarrhea. It was also taken internally to kill parasitic worms. The inner bark was also used to make a syrup that was taken as a spring tonic as well as a cough medicine.

Notes Trembling aspen is often found with a white powder covering the bark that has been used by Native peoples as protection from ultraviolet radiation. It is likely used by the tree for the same purpose!

Willow *Salix* spp.

Willow Family (Salicaceae)
Description Overall: Deciduous tree or tall shrub.
Flowers: Catkins to 4" (10 cm) long; appear before or with leaves in spring.
Fruits: Brown capsules, to 0.25" (6 mm) long. **Leaves:** Simple, deciduous; shiny green above, lance-shaped with a round base; often with a fine-toothed margin; to 5.25" (13 cm) long.
Other: Plants arise from shallow roots.
Size To 39' (12 m) high or higher.

Habitat Various wetlands including rivers, streams and lakes; sea level to mid-elevation.
Range Alaska to Mexico.
Edible Uses Aboriginal peoples occasionally ate the young shoots and leaves, buds and inner bark of willow, but they are quite bitter and considered an emergency food.
Medicinal Uses The bark of willow is world famous for containing at least 12 different salicylates. The original synthesis of acetylsalicylic acid (ASA) stemmed from willow bark.

Native peoples chewed or made medicinal teas of willow bark for treating diarrhea and other digestive problems, as well as headache, arthritis and urinary tract irritations. Bark tea or strips of softened bark were used for washes and poultices on secondary burns, insect bites, cuts, scrapes, rashes, ulcers, corns and cancers.
Precautions People who have a sensitivity to aspirin should not take willow internally. Willow contains tannin, a potential cancer-causing substance when large amounts of willow are taken internally.
Notes Numerous species of willow occur in the Northwest, and the properties vary from species to species. For example, some, like **Arctic willow *Salix artica*,** are very small, reaching a mere 2.4" (6 cm) in height. Others, such as **Pacific willow *Salix lucida lasiandra*,** tower above us to 39' (12 m). Research has shown that the compounds found in willow vary with the species and season.

Catkins.

Pacific Madrone *Arbutus menziesii*

Other Names Also known as arbutus, madrone.

Heath Family (Ericaceae)

Description Overall: Shrub, tree, perennial herb. **Flowers:** White, urn-shaped, fragrant (similar to buckwheat honey), to 0.3" (7 mm) long; in drooping clusters; appear with new leaves. **Fruits:** Berry-like, orange-red with granular surface, many-seeded, in clusters, to 0.4" (1 cm) across. **Leaves:** Simple, alternate, evergreen; shiny dark green above, whitish green below, leathery; to 6" (15 cm) long. **Other:** Plants grow from deep spreading lateral roots.

Size To 66' (20 m) high.

Habitat Dry, sunny, rocky sites; intolerant of shade; low to mid-elevation.

Range Southern BC south to Baja California.

Edible Uses Arbutus berries were eaten in small quantities by Native peoples in California.

Peelilng bark.

Medicinal Uses It is believed that diabetes was treated with arbutus bark by Cowichan groups. Other plants in the heath family have also been used to treat problems with sugar metabolism. The bark was decocted and used as a wash for eye troubles as well as minor wounds. A poultice of leaves has been used on burns and for rheumatoid arthritis and other joint problems.

Pacific madrone bark and leaves have been prepared to combat colds and stomach conditions, and as a post-childbirth contraceptive. It has also been used by some Native peoples in a ten-ingredient bark medicine used against tuberculosis and spitting up blood.

Notes The Salish people on Vancouver Island in BC used the bark to tan hides, because of its high tannic acid content.

Black Hawthorn *Crataegus douglasii*

Other Names Also known as Douglas hawthorn, western thorn apple.

Rose Family (Rosaceae)

Description Overall: Deciduous shrub with zigzag branch pattern and spines growing to 1.25" (3 cm) long. **Flowers:** White, 5 petals, in terminal clusters; strong-smelling; to 0.4" (9 mm) across. **Fruits:** Blackish purple pomes (apple-like fruit); to 0.4" (9 mm) long. **Leaves:** Alternate, leathery, oval with 5–9 lobes; to 2.4" (6 cm) long. **Other:** Plants stem from taproots.

Size To 33' (10 m) high.

Habitat Moist areas, open forest, wetland edges and similar sites; low to mid-elevation.

Range Alaska to California, east to Saskatchewan and Michigan.

Edible Uses The pomes are edible but they contain rough seeds. They are sometimes made into jams or jellies. Native peoples ate the fruit fresh but it was not considered a prime food. The fruit was also dried and used in making pemmican.

Medicinal Uses Many parts of this shrub were used for a variety of ailments. The bark was used in thinning the blood, strengthening the heart, reducing swelling and treating sexually transmitted diseases. The fruit and flowers have been used to normalize blood pressure and to treat angina and arteriosclerosis. Tea made from the hawthorn fruit is used in Chinese weight loss programs.

Precautions The seeds contain glycoside, which has digitalis-like properties, and they are poisonous if eaten in large quantities.

Notes The flowering branches are collected and dried in early spring, or in early autumn once the fruit is ripe. The pomes are picked when they are purple-black and the frost has not touched them. The dried flowers and leaves will last approximately one year, and the dried fruit will last for several years. Tinctures can be prepared with the flowering tops and/or fruit.

Similar Species Common Hawthorn *Crataegus monogyna*, which originated in Europe, grows to 50' (15 m) high with long, slender, curved thorns and leaves that resemble oak leaves. Its scarlet fruit is considered edible but somewhat astringent. The common hawthorn is an escapee; it occurs from southern BC to California. **Red Hawthorn *Crataegus columbiana*** grows to 20' (6 m) high with slender, curved thorns to 2" (5 cm) long. It produces clusters of red fruit. The red hawthorn grows from southern BC to California, east of the Cascades to Idaho.

Fruit.

Cascara *Rhamnus purshiana*

Other Names Also known as Cascara buckthorn; formerly classified as *R. purshianus, Frangula purshiana*.

Buckthorn Family (Rhamnaceae)

Description Overall: Deciduous tree or tall shrub. **Flowers:** Yellowish green, small. **Fruits:** Berry-like drupes; blackish, to 0.6" (1.4 cm) across. **Leaves:** Simple, alternate, deciduous; green above, pale green below; oval, parallel veins curving forward; tip short, base rounded; to 6.3" (16 cm) long. **Other:** Plants arise from a shallow root system.

Size Normally to 20' (6 m) high; occasionally to 33' (10 m) high.

Habitat Mainly on moist sites, burned areas, clearings and in conifer forests, where it is an understory species.

Range Southern BC to northern California; east to northern Idaho and northwest Montana.

Medicinal Uses Cascara has a long history of use as a laxative, a property that has been verified scientifically. Native peoples made a tea from its bark. Today many commercial herbal laxatives include this bark as a component, and the laxative effect of cascara is not habit-forming. The bark should be gathered and aged for one year before use, which causes its chemical structure to change to make it more effective and less nauseating. A dehydrator will hasten the process to about 12 hours. This species was also used to wash sores and swelling, and to treat heart strain.

Precautions Cascara has been known to cause severe vomiting and diarrhea. Avoid this wood if you are selecting roasting sticks for hot dogs.

Notes In past years, millions of pounds of cascara bark was harvested annually and the plant became quite rare in many areas. Governments have now regulated the harvest in an effort to stop the systematic elimination of cascara from our forests.

The distinctive leaves.

Salal *Gaultheria shallon*

Other Names Also known as laughing berries, Oregon wintergreen.

Heath Family (Ericaceae)

Description **Overall:** Evergreen deciduous shrub. **Flowers:** White to pink, urn-shaped, in elongated clusters of 5–15 at the branch ends; to 0.4" (9 mm) long; May to early June. **Fruits:** Blue to black, "berries" are actually fleshy sepals, in elongated clusters of 5–15 at the branch ends; to 0.4" (9 mm) across; July

and August. **Leaves:** Evergreen, alternate, leathery, with a shiny surface. To 4" (10 cm) long. **Other:** Plants arise from a wide, deep root system.

Size To 16' (5 m) high.

Habitat Various coniferous forests; low to mid-elevation.

Range Southeast Alaska to southern California.

Edible Uses The fruit of salal is edible. In years when it is not too dry, the berries can be very tasty, and can be combined with other berries in a wide variety of dishes. Native peoples used this berry extensively.

Medicinal Uses Salal tea is considered a mild remedy for the bothersome raspy cough of allergies or dry air, and for diarrhea when accompanied by cramps and a slight fever. The plant is classified as astringent, anti-inflammatory and carminative, and it is used as a convalescent tonic.

When young children have hypersensitivity to foods (soy milk, pea soup, whole grains, garlic) that cause colic and gas pains, salal tea has been found to be helpful. It can also soothe the symptoms of gastritis or a pre-ulcer stomach condition in adults. Salal tea can also serve as a pain reliever for minor scrapes, burns and insect bites. The

best time to gather branches of salal is late spring to mid-fall.

Notes Salal is an abundant shrub that can grow so dense that it creates a nearly impenetrable barrier. Stems and leaves of salal are often used in commercial flower arrangements.

Common Labrador Tea *Ledum groenlandicum*

Other Names Also known as swamp tea, Hudson's Bay tea, muskeg tea, Indian tea.

Heath Family (Ericaceae)

Description **Overall:** Evergreen shrub. **Flowers:** White, with protruding stamens, in short umbrella-like clusters, 5–20 per cluster. **Fruits:** Drooping cluster of dry, hairy 5-part capsules. **Leaves:** Alternate, oblong to lance-shaped, to 2.5" (6 cm) long, leathery, deep-green above with thick, rusty hairs on undersides of older leaves. **Other:** Plants arise from rhizomes.

Size To 5' (1.5 m) high; rarely over 3' (90 cm) high.

Habitat Bogs and moist coniferous forests; low to mid-elevation.

Range Alaska to northwest Oregon.

Edible Uses Common Labrador tea was valued as a beverage by European settlers and traders. (But see caution below.) Gather the leaves for tea in summer and fall, after the plant seeds, and dry them loosely in a paper bag.

Medicinal Uses Common Labrador tea has often been used to treat colds, sore throat and allergy symptoms. The tea was thought to relieve diarrhea and stomach upset, but it also acts as a mild laxative. (See caution below.) Fresh leaf tinctures have been used to treat scabies, lice and chiggers as well as fungal skin diseases. Decoctions were used to treat inflamed, itchy or oozing skin conditions.

Precautions If common Labrador tea is taken internally, it is best to do so in moderation, as excess can cause drowsiness, increased urination and intestinal disturbance. Narcotic compounds and toxins are present, which can cause cramps, delirium, palpitations, paralysis and even death. It is generally believed that common Labrador tea should be gently boiled for up to 10 minutes so that it does not release ledol, a toxin that can bring on headaches, cramps and paralysis. Boiling longer is not recommended. Labrador tea should not be used by pregnant women.

Notes The spicy fragrance of this plant makes it a favored addition to pot pourris. The leaves are also known to have a mild narcotic effect when smoked.

Similar Species **Glandular Labrador Tea *L. glandulasum*** displays pale lower leaf surfaces, with dense resin glands and short white scales. It grows in montane and subalpine habitats, and its range stretches from BC and Alberta to Wyoming. It is relatively toxic.

Common Bearberry *Arctostaphylos uva-ursi*

Other Names Also known as bearberry, kinnikinnick, pinemat manzanita, uva ursi.
Heath Family (Ericaceae)
Description Overall: Evergreen shrub, trailing, mat-forming. **Flowers:** White to pinkish, urn-shaped, drooping to 0.25" (6 mm) long, in terminal clusters. **Fruits:** Red, berry-like drupes, to 0.4" (10 mm) across. **Leaves:** Alternate, dark green and glossy above,

oval to spoon-shaped, leathery, to 1.2" (3 cm) long; dark green and shiny above, paler beneath. **Other:** Plants arise from long runners.
Size To 8" (20 cm) high.
Habitat Well-drained, open sites; foothills to alpine.
Range Alaska east to Labrador; south to Mexico.
Edible Uses The berry-like drupes are edible, but they are pulpy and not tasty.
Medicinal Uses Native peoples have used the berries and leaves of this plant to treat diarrhea. A wash for sore eyes was made from a decoction of leaves and stems, which was also taken as a remedy for kidney and bladder problems. A similar decoction that also included Oregon grape branches was used as a kidney medicine and blood tonic.
   Other uses of common bearberry include hair tonic to treat dandruff and scalp diseases, and a wash for skin sores. This species has diuretic, astringent, soothing, tonic and nephritic properties.
Precautions This species has been widely used to make tea, but it should be taken only in moderation, because it can lead to stomach and liver problems. Pregnant

women should not drink strong bearberry tea as it can bring on uterine contractions. Too many berries can also cause constipation.
Notes Common bearberry has a high tannin level and as a result has been used to tan hides.
Similar Species Hairy Manzanita *Arctostaphylos columbiana* is an evergreen shrub with hairy young leaves, and twigs and mealy, blackish red fruit. It grows to a height of 10' (3 m). Hairy manzanita is found in dry openings along the coast from southern BC to California.

Common Snowberry *Symphoricarpos albus*

Other Name Also known as waxberry.
Honeysuckle Family (Caprifoliaceae)
Description **Overall:** Deciduous shrub with hollow, hairless twigs. **Flowers:** Pinkish, bell-shaped; to 0.25" (7 mm) long; June. **Fruits:** Clusters of round, white, waxy berry-like drupes, ; to 0.6" (1.5 cm) diameter; August. **Leaves:** Simple, opposite, oval; to 0.75" (2 cm) long. **Other:** Shrub arises from rhizomes and often forms colonies.

Size To 9' (2.75 m) high.
Habitat Moist clearings, stream banks, thickets; low to mid-elevation.
Range Common throughout most of North America.
Edible Uses None, as the entire plant is **poisonous**.
Medicinal Uses Fresh leaves, fruit and bark have been used to make poultices for burns, sores, cuts, scrapes and similar mishaps. Wounds treated with common snowberry heal with very little scabbing. The root and bark have been classified as diuretic, stomachic and tonic, and the berries as astringent, cathartic and emetic.

A decoction of the roots and stems has been used to cure morning sickness. Native peoples made a twig decoction for fever and menstrual disorders. Whole berries were crushed or boiled to make a wash for sore eyes. Teething pains and fever were treated with a decoction of roots and stems.

Common snowberry was combined with another plant to make a treatment for sexually transmitted diseases. Other Native groups treated these conditions by boiling the bark and root of the common snowberry.

Precautions This plant is **toxic** throughout—do not eat the fruit or any other part of it.
Notes The berries of this plant have been rubbed into the hair as shampoo. Snowberry has also been used as one component in the making of a love potion.

Ocean Spray *Holodiscus discolor*

Other Names Also known as arrow-wood, creambush, ironwood, iron-wood.
Rose Family (Rosaceae)
Description **Overall:** Deciduous shrub. **Flowers:** White to cream, to 0.2" (5 mm) across; in terminal pyramid-like clusters 4–7" (10–17 cm) long. **Fruits:** To approx. 0.1" (2 mm) long, brown, hairy achenes. **Leaves:** Alternate, deciduous, hairy, wedge-shaped to triangular, to 0.2" (6 cm) long, coarsely toothed. **Other:** Plants arise from shallow to deep root systems, likely depending upon ground conditions.
Size To 12.5' (4 m) high.
Habitat Dry to moist open sites including open woods, clearings, ravine edges, coastal bluffs, along the coast and drier areas inland; low to mid-elevation.
Range Alaska to California.
Medicinal Uses The blossoms of ocean spray have been used to treat diarrhea. Infusions of seeds have been given to treat smallpox, black measles and chicken pox. Leaf decoctions have been made as remedies for influenza. Native peoples have applied poultices of leaves to sore lips and sore feet.

A tonic was prepared from a decoction of bark and given to convalescents and athletes to aid in their endurance. An eyewash was made from an infusion of inner bark. Dried bark was powdered and mixed with petroleum jelly to make a dressing for burns. This herb is classified as astringent, diuretic, tonic and emetic, and the leaves and flowers have been listed as especially antioxidant, cytotoxic and anti-inflammatory.
Notes The wood of ocean spray has been used by Native peoples to make arrows, knitting needles and cattail mat needles. The hardness of its older wood was used to advantage by pioneers, who used wooden wedges instead of nails in construction.

Sitka Mountain Ash *Sorbus sitchensis*

Other Name Formerly classified as *Pyrus sitchensis*.

Rose Family (Rosaceae)

Description **Overall:** Deciduous shrub. **Flowers:** White, flat-topped terminal clusters; to 0.4" (9 mm) long; June and July. **Fruits:** Red to orange or purple, whitish bloom (powdery covering); berry-like pomes; to 0.4" (9 mm) diameter; July to October. **Leaves:** Alternate; 7–11 leaflets with rounded tips, toothed along upper half of margin; to 8" (20 cm) long. **Other:** Plants arise from an extensive root system.

Size To 13' (4 m) high.

Habitat Open coniferous woods, stream banks, parklands; mid- to subalpine elevation.

Range Alaska and southern Yukon to northern California.

Edible Uses The Okanagon and Thompson Native peoples occasionally used the fruits for food, probably cooked.

Medicinal Uses The fresh fruit was crushed and eaten raw for its strong purgative properties. The bark of the roots, and occasionally the inner bark of the stem, was boiled for an hour, and the decoction taken internally for stomach problems or rheumatoid arthritis. This tea was also used externally, as an eyewash and as a bath for people with rheumatoid arthritis. The bark was also chewed to relieve the symptoms of colds. The peeled branches or inner bark was boiled to make a tea that was used in treating colds, headache and chest and back pain. The Bella Coola and other Native people rubbed the mashed berries into the scalp as an insecticide for lice.

Precautions The seeds of this fruit contain cyanide and are poisonous. Do not eat them!

Notes The leaves of the Sitka mountain-ash make it an easy plant to find in autumn, when they display spectacular colors in contrast to the surrounding species. The berries are rich in vitamin C.

Wild Rose *Rosa* spp.

Rose Family (Rosaceae)

Description **Overall:** Thorny shrub. **Flowers:** Pink to deep rose; to 3" (8 cm) across; normally found at tips of branches. **Fruits:** Red, round to elongated hips, to 0.8" (2 cm) across, with numerous hairy achenes; ripe in late August to October. **Leaves:** Odd number of leaflets (normally 5 or 7); large thorns may be present at the base of each leaf. **Other:** Plants arise from deep taproots.

Size To 10' (3 m) high.

Habitat Open areas including shorelines, meadows, stream banks, roadsides, clearings; low to mid-elevation.

Range Alaska to México, including most of the continent.

Nootka rose

Edible Uses All parts of the wild rose were consumed by various Native peoples. An acceptable potherb was made from the young shoots of spring. The stems and roots were brewed to make tea. Fresh flower petals were added to salads for flavor and color. The rosehip is the best-known edible part of the wild rose. Only the outer rind of the fruit was utilized, because the seeds contain hairs that are very irritating to the digestive tract. The hip is said to become sweeter after being touched by frost. Rosehip jam can be made, but it is rather bland. Use only the outer rind (see below), and combine it with other fruits such as saskatoons or blueberries to give it much more flavor. Rosehips can also be made into tea or syrup.

Medicinal Uses Native peoples of North America were using wild roses before the beginning of recorded history. The flowers were soaked in rainwater and used to bathe sore eyes. The petals were eaten to help reduce fever. A tea made from the petals was combined with mint and taken internally for relaxation and sleep. A root decoction was administered internally to treat menstrual irregularity. Rose petals were also used as a heart tonic.

Precautions The seeds and hairs in rosehips are always removed, because if swallowed they can irritate the intestinal tract. The seeds also contain cyanide-like compounds—another reason they must not be eaten.

Notes The wild rose is well known for its beautiful flowers, wonderful scent and bristly branches. Many species may be encountered in this region (a few are listed below). Three rosehips contain as much vitamin C as an orange. Note, however, that the vitamin C content of rosehips varies widely with the species.

Similar Species **Nootka Rose** *Rosa nutkana* is a coastal species found west of the Cascades.

Baldhip Rose *Rosa gymnocarpa* drops its sepals early, so they are not found on this plant's rosehips.

Clustered Wild Rose *Rosa pisocarpa* is a native species with clusters of beautiful, small flowers.

Shrubby Cinquefoil *Potentilla fruticosa*

Other Names Also known as snow cinquefoil, *Pentaphylloides floribunda*.
Rose Family (Rosaceae)
Description Overall: Deciduous shrub, often with shredding, reddish bark when mature. **Flowers:** Yellow, buttercup-shaped, to 1.2" (3 cm) across, with 5 broad petals, often in small clusters at branch tips. **Fruits:** Light brown achenes, densely covered with long white hairs. **Leaves:** Alternate, pinnately compound, with 3–7 (normally 5) crowded leathery leaflets, to 0.8" (2 cm) long, silky. **Other:** Plants arise from fibrous roots.
Size To 4.9' (1.5 m) high.
Habitat Wet to dry sites including boggy areas, rocky sites, meadows and rocky slopes; plains to subalpine.
Range Alaska to New Mexico.
Edible Uses The young leaves can be dried and made into an excellent golden tea that is rich in calcium.

Medicinal Uses The leaves, stems and roots have been boiled together and the resulting liquid taken internally to treat a fever with body aches. The entire herb is dried. The leaves were used for weak bowels, internal hemorrhage, influenza, stomach problems and other ailments. Stomach and esophagus inflammations have been treated with this tea 3 to 4 times a day.

Notes Tea made from shrubby cinquefoil was also used to prevent saddle sores on horses. The animal was soaked with the tea.

An increase in shrubby cinquefoil is an indicator of overgrazing by wild or domestic animals, and is often observed as part of controlling the numbers of animals in a grazing area.

Shrubby cinquefoil has been cultivated, and several varieties are available as shrubs for the garden.

Devil's Club
Oplopanax horridus

Other Names Also known as devil's-club; formerly classified as *Echinopanax horridum*.

Ginseng Family (Araliaceae)

Description Overall: Deciduous shrub armed with numerous large spines to 0.4" (9 mm) long. **Flowers:** Greenish to white, many on an upright raceme; raceme to 0.75" (2 cm) long; May to July. **Fruits:** Showy red, flattened clusters in a pyramidal raceme; to 0.3" (8 mm) long; July to September. **Leaves:** Alternate, maple leaf-shaped with 5–7 lobes; to 14" (35 cm). **Other:** Plant grows from spreading rootstocks.

Size To 10' (3 m) high.

Habitat Moist, open areas such as forest edges and stream banks; low to mid-elevation and to timberline in the north.

Range Alaska to Yukon Territory, south to Oregon and east to Idaho and Montana.

Medicinal Uses Devil's club is generally regarded as safe and reliable to ingest as an expectorant. Various parts of devil's club were used historically by many Native peoples to treat maladies. Several pieces of stem, with the spines scraped off, were infused in water and drunk to relieve symptoms of arthritis and rheumatoid arthritis. The bark was dried and made into an infusion to treat colds and to be applied topically for rheumatoid arthritis by the Cowlitz. A poultice was made to treat rheumatoid arthritis and pain by the Cowichan, Sechelt and Squamish. The stems were steeped and the tea was drunk to treat colds. The root and bark were steeped by the Skagit and mixed with prince's pine and cascara as a treatment for tuberculosis. The fruit was rubbed into the scalp to combat lice and dandruff. The inner bark was used as a cure for rheumatoid arthritis and tuberculosis of the bone. The stems and branches were boiled to make a tea that was drunk to reduce fever. The inner bark of the root was used in treating tuberculosis, stomach trouble, fever, cough, colds, swollen glands and boils.

Devil's club also has a long history of use by Native peoples for adult-onset, insulin-resistant diabetes, and its benefits for this condition have been confirmed in several studies.

The well-known herbalist Michael Moore suggests collecting the fresh bark of the stem-roots and the bark and the true root heartwood in late summer and early fall for a tincture or tea. Another prominent herbalist, Terry Willard, prefers to collect the spring bark for his tinctures.

Precautions Caution is advised in handling devil's club. Its spines can break off under the skin and produce a surprisingly nasty and slow-healing abrasion.

Notes Devil's club is probably best known for its value in treating diabetes. Despite its intimidating appearance and its painful spines, it is truly a remarkable plant!

Red Osier Dogwood *Cornus sericea* var. *occidentalis*

Other Names Also known as red-osier dogwood, red willow; formerly classified as *C. alba*, *C. stolonifera* and *Svida sericea*.

Dogwood Family (Cornaceae)

Description Overall: Deciduous shrub. **Flowers:** White, to 0.2" (5 mm) across, in flat-topped clusters, to 2" (5 cm) across at the tips of branches; May to July. **Fruits:** White berry-like drupes with flattened stones, to 0.3" (7 mm) across. **Leaves:** Opposite, oval to lance-shaped, to 4" (10 cm) long, pointed, parallel veins converging to the tip. **Other:** Plants arise from a thick, fibrous root system and stolons (creeping stems that root at the nodes).

Size To 10' (3 m) high.

Habitat Moist, wooded areas, along streams, open sites; plains to montane areas.

Range Central Alaska south to California and east to New Mexico.

Edible Uses The white berry-like drupes were often collected for food by Native peoples in the past. They are quite bitter but they ripen at the same time as chokecherries. The two fruits were often mixed together and pounded to make them more palatable. The fruit of red osier dogwood contains 30 percent more energy than other fruits because lipids (fats) are present. (See caution below.)

Medicinal Uses The bark of this plant contains comic acid, a compound similar to but milder than the familiar acetylsalicylic acid (ASA). A decoction of the inner and outer bark was sometimes used as a cough syrup. Other dogwood species have had similar uses for cough suppression and treatment of infections and lung maladies.

Some Native groups used the roots as a remedy for diarrhea and as a mild astringent.

Precautions It is now known that all parts of red osier dogwood can be toxic, especially when large quantities are consumed.

Notes The bark of red osier dogwood was made into a type of rope by twisting it, then making it more supple by rubbing it as it dried. This rope was used for lashing fish traps and raised caches, and similar needs.

Beaked Hazelnut *Corylus cornuta*

Other Names Also known as California hazelnut, hazelnut; formerly classified as *Corylus californica*.

Birch Family (Betulaceae)

Description Overall: A tall, many-branched deciduous shrub with smooth grayish brown bark. **Flowers:** Male flowers in catkins; female catkins very small and inconspicuous, with protruding red stigmas. **Fruits:** A spherical nut is surrounded by a tube of fused, stiff and prickly bracts, which project forward into a beak-like shape up to 1.2" (3 cm) beyond the nut. Clusters of 2 or 3 nuts are present at the branch ends. **Leaves:** Simple; oval and double-toothed.

Size To 10' (3 m) high.

Habitat Moist, well-drained sites in a wide variety of situations including open forest, shady openings, clearings, rocky slopes, stream areas; low to mid-elevation.

Range BC to the east coast; south to California.

Edible Uses These edible nut were picked in the early autumn before squirrels and other small mammals began to harvest them, and stored until fully ripe. The prickly husks must be removed prior to eating. The nuts can then be stored for extended periods and eaten as they are. They can also be ground into flour, but the flour cannot be stored very long. Native peoples husked hazelnuts by various means, including pounding them with a pole.

Notes The nut of a closely related species, the American hazelnut *C. americana*, found farther south, contains high values of energy. This is likely true of beaked hazelnut as well. Native peoples produced various dyes from the beaked hazelnut. The nut can yield a green dye if boiled for 30 to 60 minutes, and the bark was boiled to produce a reddish brown dye.

Canada Buffaloberry *Shepherdia canadensis*

Other Names Also known as soapberry, Canadian buffalo berry, soopolallie.
Oleaster Family (Elaeagnaceae)
Description Overall: Deciduous shrub. **Flowers:** Greenish, tiny, to 0.2" (4 mm) across, male or female flowers on separate plants, in small clusters below new leaves; April to June. **Fruits:** Bright, transparent, red to orange berries, to 0.25" (6 mm) across. **Leaves:** Opposite, oval, to 0.25" (6 cm) long, dark green above and fuzzy below with rust-colored scales. **Other:** Plants arise from a shallow root system.
Size To 6.6' (2 m) high.
Habitat Open woods and stream banks.
Range Alaska east to the east coast, south to California.
Edible Uses Canada buffaloberry fruit was a special treat for Native peoples, referred to by newcomers as "Indian ice cream." To prepare it they would crush the fresh fruit with their fingers and remove the seeds, then whip the rather bitter fruit into a froth. Saskatoon berries or strawberries might be added to sweeten it. Later on, it was sweetened with sugar. The fruit was also dried in cake form for future use.
Medicinal Uses The leaves and stems have been boiled to make a decoction used in treating tuberculosis. This decoction has also been used as a wash for cuts, swellings, arthritis and other troubles with aching limbs or joints. New shoot tea has been drunk to prevent miscarriages, to treat sexually transmitted diseases and coughing up blood, and as a wash for easing arthritis.
Precautions Although the berries are rich in vitamin C and iron, large amounts are known to cause cramps, diarrhea and vomiting due to the saponin, a soap-like substance, that it contains.
Notes Canada buffaloberry is often called soapberry because of the froth that is produced when the berries are whipped.
Similar Species Thorny Buffaloberry *Shepherdia argentea* is a thorny species whose range extends from southern BC to California.

Tall Oregon-grape *Mahonia aquifolium*

Other Names Also known as mountain grape, dragon grape, mountain holly, mahonia, pepperidge, barberry, American barberry, European barberry, California barberry, common barberry, jaundice barberry, blue barberry, creeping barberry, berberry, sourberry, wood sour, sowberry, piprage, algerita, guild tree, japonica, yellow root, pepperidge bush; formerly classified as *Berberis aquifolia*.

Barberry Family (Berberidaceae)

Description Overall: Evergreen shrub. **Flowers:** Yellow; to 0.5" (1 cm) long; in erect clusters. **Fruits:** Blue, whitish bloom or powder, grape-like berries; to 0.35" (9 mm) diameter, August and September. **Leaves:** 5–11 leaflets, very glossy above; to 3" (7.5 cm) long. **Other:** Plants arise from taproots.

Size To 5' (1.5 m) high.

Habitat In dry, open forests.

Range Southern BC south to northern California, west to Idaho.

Edible Uses The berries of this species are quite juicy and are often used to make pies and jellies. To make jelly, boil the berries with an equal measure (or a little less) of sugar, then strain.

Medicinal Uses Tall Oregon-grape is one of our most valuable herbs. It is considered a tonic, antiseptic, mild laxative, stimulant, bitter and refrigerant. The roots are normally collected and dried, but the leaves have been used as well. Crushed plants and root tea were used to aid healing of wounds, as they are antiseptic and antibacterial. The leaf tea has been taken as an overall tonic and as a contraceptive. In addition, kidney troubles, stomach troubles, rheumatoid arthritis and loss of appetite were treated with tall Oregon-grape. The bark of the root is considered the most potent—three times stronger in alkaloids than the bark on the stem.

Precautions The National Standard Dispensatory lists many uses, but it also warns that an overdose can be fatal! All parts of the Oregon-grape contain berberine, an alkaloid drug with antibiotic and analgesic properties. The roots contain the highest concentration of berberine, which is very potent and potentially toxic.

Notes Tall Oregon-grape is the state flower of Oregon. The closely related species below are similar to tall Oregon-grape in edible and medicinal properties.

Similar Species Dull Oregon-grape *Mahonia nervosa* is a smaller species that grows to 2' (60 cm) tall, with 9–19 leaflets. These leaflets have 3 main veins and a dull appearance, unlike the glossy tall Oregon-grape. Dull Oregon-grape grows along the coast from BC south to California.

Creeping Oregon-grape *Mahonia repens* grows to a height of 2' (60 cm), with 3–7 leaflets on each compound leaf. The teeth on the leaves are much shorter than those of tall Oregon-grape. This species favors dry sites east of the Cascades, from BC south to New Mexico, including the Great Plains.

False Solomon's Seal *Maianthemum racemosum*

Other Names Also known as false spikenard; formerly classified as *Smilacina racemosa, Vagnera racemosa*.

Lily Family (Liliaceae)

Description **Overall:** Perennial herb. **Flowers:** Creamy white, to 0.1" (3 mm) long, forming a large, puffy terminal pyramidal cluster. **Fruits:** Reddish, round berries, often spotted with purple; to 0.3" (7 mm) diameter. **Leaves:** Alternate, elliptical, parallel veins; to 8" (20 cm) long. **Other:** Plants arise from stout, fleshy rhizomes.

Size To 40" (1 m) high.

Habitat Moist woods, stream banks; low to subalpine.

Range Alaska to southern California.

Edible Uses The Lakes Salish people pulled up the rhizomes, washed them and ate them uncooked (they taste similar to onions), or cooked. The Blood people used the leaves in their soups and made the rhizomes into pickles. The berries were also eaten raw, but they are purgative if eaten in large quantities.

Medicinal Uses Blackfoot people dried and powdered the rhizomes and applied the substance to wounds. The Blood people used the rhizomes to induce abortion. The rhizomes have also been boiled in water to make a medicine that relieved symptoms of colds or increased the appetite.

Precautions The berries may cause diarrhea if they are eaten in large quantities.

Notes The leaves of false Solomon's seal were made into a tea that was ingested daily by Native women of Nevada as a form of birth control. The plant contains a small amount of a steroid, diosgenin, which at one time was used as the starting material for a synthetic birth control hormone. The berries are high in vitamin C.

Similar Species **Star-flowered Solomon's Seal** *Maianthemum stellatum* is a similar species that flowers with star-like terminal clusters. Its fruits, which are greenish yellow striped with purple, mature to dark blue or black. The roots of this closely related species were made into infusions as a form of birth control.

Nodding Onion *Allium cernuum*

Description **Overall:** Perennial herb. **Flowers:** Pink or white flowers throughout the summer months. **Fruits:** Capsules with 6 points at tip, to 0.2" (4 mm) long. **Leaves:** Hollow and round in the cross-section, with the distinct aroma of onion. **Other:** Short rhizomes with egg-shaped bulbs.

Size To 20" (50 cm) high.

Habitat Present in a wide range of habitats including subalpine meadow, parkland, prairie, open slopes and thickets.

Range BC to New Mexico.

Edible Uses The bulbs were commonly eaten by Native peoples and European settlers, either raw, cooked or dried and stored for winter. The leaves can also be eaten raw, or added to a salad.

Medicinal Uses Onions are antibacterial, antiviral and antifungal. They are also reputed to be a

stimulant, carminative, antiseptic and diuretic. The crushed green leaves were applied

directly to scalds, burns and wounds to prevent infection, as well as to sores. Onions are also said to help digestion.

Precautions It is imperative that the onions chosen are true onions rather than the similar looking and extremely poisonous meadow death-camas (see p. 75), which is deadly. A taste test is NOT the way to determine whether you have an onion. The smell is distinctive for all onions. If your specimen does not smell like an onion, discard it.

Common Camas *Camassia quamash*

Other Names Also known as blue camas, early camas, quamash.

Lily Family (Liliaceae)

Description Overall: Perennial herb. **Flowers:** Pale to deep blue, to 1.4" (3.5 cm) long; 5 or more in a terminal spike. **Fruits:** Egg-shaped capsules to 1" (2.5 cm) long; stalk curved in toward stem. **Leaves:** Numerous, basal, grass-like, to 20" (50 cm) long. **Other:** Plants arise from an egg-shaped bulb, 0.8" (2 cm) long.

Size To 27.5" (70 cm) high.

Habitat Grassy slopes and meadows; low to mid-elevation.

Range Southeast Vancouver Island, BC, to California; scattered records elsewhere.

Edible Uses The bulbs of common camas were collected by Native peoples as an important food item. The best meadows were fought over by groups trying to obtain their choice root crops. Only large bulbs were gathered, leaving smaller ones for harvesting in future years. The bulbs were then cooked very slowly in large pits for 24 hours or more to break down the large carbohydrates into more digestible sugars. The cooked bulbs were ground and formed into thin cakes that were dried in the sun and then stored.

Precautions Be sure that meadow death-camas (see below) is not included with your harvest. As its name suggests, it is highly toxic and potentially fatal.

Notes Growing areas of common camas were carefully tended by the Vancouver Island Coast Salish people. The plant was semi-cultivated—the area was cleared of stones, weeds and brush, and underwent controlled burning in some summers. These areas could be inherited. This species is less common now that urban and other developments are overtaking many of the areas where it was abundant.

Similar Species Meadow Death-camas *Toxicoscordion venenosum* (see p. 75) displays distinctive white to cream-colored flowers and a similar-looking bulb.

Broad-leaved Stonecrop *Sedum spathulifolium*

Other Name Broadleaf stonecrop.
Stonecrop Family (Crassulaceae)
Description Overall: Perennial herb. **Flowers:** Bright yellow with 5 lance-shaped petals, to 0.4" (1 cm) long, in clusters. **Fruits:** Star-shaped clusters above. **Leaves:** Alternate, green to red, flattened and wider at the tips. **Other:** Plants arise from rhizomes.
Size To 8" (20 cm) high.
Habitat Cliffs and rocky areas; low to mid-elevation.
Range Alaska to California.
Edible Uses The young leaves and shoots of all stonecrops are edible, but should be eaten only in moderation (see below).
Medicinal Uses Patients with fevers were said to receive cooling effects from the bruised leaves of stonecrop. Some species of stonecrop have been used as medicine to eliminate worms and to help with warts. The plant was used externally to relieve the pain and soreness of cankers and to help with piles. Another species of stonecrop has been used as a cure for sexually transmitted disease, by steeping the whole plant in a small amount of hot water and drinking the liquid.
Precautions The leaves of all stonecrops are edible, but over-indulgence is not recommended. Some have emetic (vomiting) and cathartic (laxative) properties and can cause headaches.
Notes Broad-leaved stonecrop is one of several species of colorful stonecrops growing in the Pacific Northwest (see below). Some varieties can be found in gardens.
Similar Species
Oregon Stonecrop *S. oreganum* displays fleshy, spoon-shaped leaves reminiscent of a succulent.
Lance-leaved Stonecrop *S. lanceolatum* leaves are rounded and lanceolate.

41

Roseroot *Sedum integrifolium*

Other Names Also known as king's crown; formerly classified as *S. rosea, S. roseum, Tolmachevia integrifolia.*

Stonecrop Family (Crassulaceae)

Description **Overall:** Succulent perennial herb. **Flowers:** Dark purple to pink, clustered at top of plant. **Fruits:** Red to purplish follicles, star-like clusters of 5 capsules with pointed tips. **Leaves:** Green or pink, fleshy, oval, alternate, to 1.6" (4 cm) long. **Other:** Plants arise from rhizomes.

Size To 8" (20 cm) high.

Habitat Moist, rocky slopes; subalpine to alpine.

Range Alaska to California; east to Colorado.

Edible Uses Leaves are succulent and juicy. They can be eaten raw in a salad or cooked as a potherb. Roseroot leaves, rich in vitamins A and C, are considered a good emergency food. The rhizomes can be boiled and eaten alone, or mixed with other vegetables. This species turns bitter and fibrous in late summer and should be collected before it flowers.

Medicinal Uses Infusions of roseroot leaves and root decoctions were made and administered for colds and gargled for sore throats. Eye irritations were also treated with these substances, by the Dena'ina Athabascans.

Notes The rhizome of roseroot emits the fragrance of roses when it is cut or bruised, hence its common name.

Blue Clematis *Clematis columbiana*

Other Names Also known as blue virgin's bower, Columbia bower, Columbia clematis, purple clematis, western blue clematis, western blue virginsbower; formerly classified as *C. occidentalis, C. verticillaris*.
Buttercup Family (Ranunculaceae)
Description Overall: Perennial vine. **Flowers:** Blue to purplish, with 4 spreading petal-like sepals; to 2.4" (6 cm) long; May to July. **Fruits:** Small achenes with long, feathery bristles in clusters. **Leaves:** Opposite, compound with 3 oval to lance-shaped leaflets; leaflets to 2.4" (6 cm) long. **Other:** Plants arise from deep roots.
Size To 16' (5 m) high.
Habitat Moist to dry sites in open woods or rocky areas, scattered and common; low to mid-elevation.
Range BC (east of the Cascades) east to Alberta and Utah; south to Baja California, Mexico.
Medicinal Uses This tea has been used as a poultice for sores, itchy skin and leg ulcers. A tincture of the plant is occasionally used as a counter-irritant. Blue clematis has also been used to treat migraines. Fresh poultice has been used to relieve painful joints. The whole vine is utilized while in leaf.
Precautions Some individuals have a skin sensitivity to this plant and others can develop swollen, inflamed eyelids. Protoam, a compound found in blue clematis, is a poisonous agent that causes respiratory paralysis. This plant should not be taken internally.
Notes Blue clematis is easily found in the forest, distinctive with its twisting, semi-woody vine that grasps and climbs the surrounding vegetation. In autumn its frilly achenes are distinctive.

Achenes.

Silverweed *Potentilla anserina*

Other Names Also known as silver-weed; formerly classified as *Argentina anserina*.
Rose Family (Rosaceae)
Description **Overall:** Perennial herb. **Flowers:** Yellow, with 5 broad petals, to 0.8" (2 cm) across, non-waxy, single on leafless stalks. **Fruits:** Produce achenes, in dense clusters surrounded by an enlarged calyx. **Leaves:** Pinnately compound. **Other:** Plants arise from thick, fleshy roots and runners that root and form leaf clusters at the nodes.
Size To 16" (40 cm) high.
Habitat Pond edges, stream edges, estuaries, beaches, dunes and salt marshes.
Range Yukon and NWT to New Mexico.
Edible Uses The roots can be eaten raw, but usually they are roasted, boiled or fried to remove the bitter taste. The cooked roots are said to taste like parsnips, sweet potatoes or chestnuts. Most Native peoples ate this plant regularly. The ideal time to dig the roots is late fall or early spring.
Medicinal Uses The Haida boiled the roots to make a tea that was used as a purgative. The roots have also been mixed with other herbs to make various medicinal preparations. A poultice made of the boiled root and fish oil was used by the Kwakwaka'wakw people. They also used the juice from the roots to treat inflamed eyes.
Precautions This species is known to stimulate the uterine muscle, so pregnant women are advised not use it.
Notes The Kwakwaka'wakw people of BC often cooked silverweed roots at their feasts. Chiefs and other high-ranking individuals would eat the choicest roots, leaving the smaller roots for the others.

Red Clover *Trifolium pratense*

Other Names Also known as purple clover, peavine clover, cowgrass.

Pea Family (Fabaceae)

Description Overall: Short-lived perennial herb. **Flowers:** Red to deep pink, sepals fused into a tube and clustered into a dense head (racemes) to 1.2" (3 cm) wide; June to September. **Fruits:** Small pods, each containing 2 seeds. **Leaves:** 3 leaflets, immediately below flowers; often a crescent-shaped spot near the base.

Other: Plants arise from woody taproots.

Size To 32" (80 cm) high.

Habitat Common and widespread in fields, grassy areas and on disturbed soils; low to mid-elevation.

Range Central Alaska to California.

Edible Uses Some Native peoples ate the tender young leaves raw of red clover—a practice that should, however, only be done sparingly (see below). The leaves of red clover contains a high volume of protein. The flowers have been boiled and eaten as well.

Medicinal Uses A tea of red clover, brewed from the flowers, has been taken to improve sluggish appetites, regulate digestive functions and treat liver problems. The flowers were also taken internally to cleanse the system, and are still taken internally as medicine for coughs and bronchitis. Red clover is listed as safe by the US Food and Drug Administration.

Externally, red clover tea has been used as a treatment for rheumatic and gout pain. A strong tea of red and white clover flowers (see below) was steeped from them individually and applied externally to boils, ulcers and other skin ailments. Red clover flowers are also used externally in British herbal therapy today, to treat psoriasis, eczema and rashes. It has also been used as a poultice for athlete's foot and other problems of the skin.

Precautions The leaves and stems of several species of clover contain chemicals that release cyanide. Therefore, use these parts with great caution and moderation.

Notes Red clover is a European species that is now widely established as a ground cover throughout most of North America and commonly used as a forage for cattle.

Similar Species White Clover *T. repens* sprawls low on the ground.
Alsike Clover *T. hybridum* stands erect with light pink flower heads.

Fireweed *Epilobium angustifolium*

Other Names Also known as bay willow-herb, bay willow, blooming Sally, burntweed, deerhorn, firetop, flowering willow, French willow herb, grand willow-herb, herb wickopy, Indian wickup, Persian willow, pilewort, Sally-bloom, spiked willow-herb, wild asparagus, willow-weed; also classified as *Chamamerion angustifolium*.

Evening Primrose Family (Onagraceae)

Description Overall: Perennial herb. **Flowers:** Pink to purple, saucer-like shape, with 4 broad petals and 8 large anthers, to 1.6" (4 cm) across; many in a single elongated cluster on top of stem. **Fruits:** Pod-like capsules to 3.5" (9 cm) long, green to red, 4-chambered, splitting lengthwise to release hundreds of seeds, each tipped with a fluffy, white tuft of hairs. **Leaves:** Alternate, green in summer, red in fall, lance-shaped, to 8" (20 cm) long. **Other:** Plants arise from spreading rhizomes.

Size To 10' (3 m) high.

Habitat In open sites, including woods and recently burned areas, and along roadsides.

Range Aleutian Islands, Alaska to Oregon.

Edible Uses Fireweed leaves, high in vitamins A and C, can be eaten raw or cooked. The flowers add a zest of color and taste to a fresh salad. The cooked leaves have been considered a good green potherb. Dried mature leaves are used to make fireweed tea, which is enjoyed around the world.

Medicinal Uses The roots and leaves have been used in treating diarrhea, eczema and sore throat. The roots have been made into a poultice or ointment that has been used to treat boils, ulcers, rashes and other skin disorders. This plant has also been reported to relieve the pain of diseased mucous membranes, colon troubles, cholera and dysentery. Firweed is considered antiscorbutic, astringent, antispasmodic, demulcent and emmenagogue.

Precautions Some individuals have found this species to have a slight laxative effect.

Notes Fireweed is a spectacular common wildflower that is well known as a colonizer of burned areas. It spreads rapidly, changing a blackened landscape into an amazing sea of pink within just a few months.

Wild Sarsaparilla *Aralia nudicaulis*

Ginseng Family (Araliaceae)

Description Overall: Woody perennial. **Flowers:** Greenish white umbels or ball-shaped clusters, each flower with 5 petals; to 0.1" (3 mm) long; May. **Fruits:** Greenish white changing to dark purple or black when ripe, forming a drupe or ball-shaped cluster; to 0.25" (6 mm) diameter, July. **Leaves:** Single and compound with 3 major divisions, each of which may have 3–5 leaflets, toothed edge; to 20" (50 cm) long. **Other:** Plants arise from a rhizome base.

Size To 2' (60 cm) high.

Habitat Shaded forests, especially mixed-wood stands; low elevation.

Range Across Canada to northeastern Washington, Montana, Colorado, Michigan and the eastern US.

Flowering plant.

Edible Uses The reddish roots (rhizomes) are sweet and were cleaned, mixed with oil and eaten by Native peoples. They were primarily considered to be an emergency food. The fruit has also been used to make beer and wine.

Medicinal Uses Wild sarsaparilla has been used in treating a wide range of ailments in-

cluding rheumatoid arthritis, gout, skin eruptions, ringworm, scrofula, internal inflammation, colds, catarrh, fever, stomach trouble and intestinal gas. It has also been used as an antidote to deadly poisons, especially when mixed with burdock (*Arctium minus* & spp.) and administered as a drink.

Notes Wild sarsaparilla often forms an impressive understory layer in forests of various types. The fruit is produced sporadically and was used as a flavoring by early North American settlers. This species got its name from a similar-tasting but unrelated tropical species, prickly vine sarsaparilla (*Smilax officinalis*). Native people in the Bella Coola, BC, area made a drink by boiling the rhizomes of wild sarsaparilla in water.

Cow-parsnip *Heracleum maximum*

The large palmate leaves are not visible here.

Other Names Also known as Indian celery, Indian rhubarb; formerly classified as *Heracleum lanatum, H. sphondylium*.
Carrot Family (Apiaceae)
Description Overall: Hairy perennial herb with hollow stems. **Flowers:** Small and white, shaped into umbrella-like clusters (umbels), to 12" (30 cm) across. **Fruits:** Flattened, with a pair of 1-seeded halves, egg- to heart-shaped, to 0.5" (12 mm) long, with vertical ribs and 2 broad wings. **Leaves:** Very large to 12" (30 cm) wide; with palmate lobes. **Other:** Plants arise from a stout taproot or clusters of fibrous roots.
Size To 10' (3 m) high.
Habitat Open meadows, moist slopes and forest edges; prairie to subalpine.
Range Alaska to California; east to New Mexico.
Edible Uses The flower stems and leaf stalks were collected and peeled, and the center portions eaten fresh. They were said to taste somewhat like celery—a likely reason for its alternate common name, Indian celery. Young stems were also peeled, roasted and eaten. The roots were eaten fresh, or boiled like potatoes.
Medicinal Uses Toothache could be relieved by placing fresh or dried root against the tooth. The root was used with other ingredients in making a medicinal tea for treating cancer. The root (fresh or dried) has been used in various ways to treat swollen legs, sore and aching body parts, and arthritis. The roots were also mixed with other herbs in the form of a poultice and used to treat aching limbs.

A fresh seed tincture has been applied topically to relieve toothache and gum abscess, as with clove oil. A fresh root tincture was used to stimulate nerve growth after suffering an injury. The seeds were also used to treat severe headaches by the Meskwaki peoples.
Precautions Be sure you know this plant before you harvest it. Several similar-looking species, also present in this area, are extremely poisonous! (See below.)
Cow-parsnip contains furanocoumarins, which can cause some people's skin to develop dark blotches, rashes and blisters on exposure to sunlight. These chemicals have been shown to cause cancer and to trigger dangerous cell alterations in some animals.
Notes The large, palmate, somewhat maple leaf-shaped leaves of cow-parsnip make it easy to identify when compared with similar-looking poisonous species. A few people have sensitivity to this plant, so approach it with caution.
Similar Species Douglas's Water-hemlock *Cicuta douglasii* has oblong to lance-shaped leaves, divided in threes with toothed leaflets. The stems are chambered at the base. **Poisonous!**
Poison-hemlock *Conium maculatum* has leaves that are dissected and fern-like, giving the plant a lacy look. **Poisonous!**

Single Delight *Moneses uniflora*

Other Names Also known as one-flowered wintergreen, shy maiden, wax-flower, wood nymph; formerly classified as *Pyrola uniflora*.

Wintergreen Family (Pyrolaceae)

Description **Overall:** Evergreen perennial herb. **Flowers:** White to greenish, solitary, waxy, saucer-shaped, to 0.6" (1.5 cm) across with 5 spreading petals. **Fruits:** Spherical, capsules split into 5 parts when mature; numerous tiny seeds. **Leaves:** Basal, oval, shiny green and finely toothed. **Other:** Plants arise from very slender creeping rhizomes.

Size To 4" (10 cm) high.

Habitat In moist forests; low and subalpine elevation.

Range Alaska to California; east to the Rocky Mountains.

Edible Uses The fruits were eaten by the Native peoples of Montana.

Medicinal Uses Single delight has been used as a remedy for a wide variety of illnesses. The plant has been used to draw blisters. A tea was made from the vegetative parts (excluding the flowering or fruiting stems) and taken for colds and flu, smallpox and cancer. The decoction is reported to be useful for sore eyes. Infusions of dried plants have been used for coughs and colds, and as a medicine for paralysis.

Leaf poultices have been applied to boils or abscesses to draw out the pus. Pounded plant poultices (chewed) have been applied to swellings, blisters and painful areas. The Haisla and Hanaksiala chewed this plant for sore throats. The Haida found that the best time to collect and dry these plants was in July.

Notes These fragrant, nodding blossoms are a wonderful sight on the forest floor. If you harvest this plant, as with any other, be sure to take only a small amount, so that the plant will remain in the forest for years to come.

One-sided Wintergreen *Orthilia secunda*

Other Names Also known as nodding wintergreen, sidebells, sidebells wintergreen; formerly classified as *Pyrola secunda*.
Wintergreen Family (Pyrolaceae)
Description Overall: Evergreen perennial herb. **Flowers:** White to pale green, bell-shaped, to 0.24" (6 mm) across; 4–20, all on one side of the stem, in long clusters (racemes); May to August. **Fruits:** A dry capsule. **Leaves:** Near base, alternate on lower section of stem, evergreen; blades longer than stalks, elliptic, to 1.6" (4 cm) long, finely toothed. **Other:** Plants arise from slender, branched rhizomes.
Size To 8" (20 cm) high.
Habitat Moist to dry sites in woods and thickets; low to subalpine elevation.
Range Alaska to Mexico.
Medicinal Uses Herbalists recommend the use of one-sided wintergreen in treating various gynecological disorders and inflammations. This species has been used for ailments including sterility, bleeding, infantilism, cervical erosion, menstrual-cycle derangements and toxicosis, and is a good anti-inflammatory agent. Elsewhere in the northern hemisphere it has been used as a diuretic and antiseptic for kidney and bladder inflammations. A strong root decoction of one-sided wintergreen has been used as an eye wash. The leaves have been chewed to relieve toothache and mixed with lard to make a salve applied to cuts to stop bleeding and promote healing.

The leaves of several wintergreens contain acids that help to heal skin problems including pimples and rashes. The leaves of several species of wintergreen have also been mashed into a poultice and used as a salve in the treatment of snake and insect bites.
Notes The dainty one-sided wintergreen is a small forest plant that finds its way to other locations, as high as the subalpine.

Field Mint *Mentha arvensis*

Other Names Also known as rook mint, Canada mint, wild mint; also referred to as *Mentha canadensis*.

Mint Family (Lamiaceae)

Description **Overall:** Aromatic perennial. **Flowers:** White to pink or pale purple, to 0.3" (7 mm) long; tube-shaped with 4 spreading lobes; numerous flowers in compact whorls from middle to upper leafaxils. **Fruits:** The 4 nutlets are enclosed by the sepals. **Leaves:** Opposite, lance-shaped to oval, to 3" (8 cm) long, short-stalked, with saw-toothed margins. **Other:** Squarish stems grow from creeping rhizomes.

Size To 31" (80 cm) high.

Habitat Wet areas such as stream banks, wet meadows, clearings, seepages, lakeshores; low to mid-elevation.

Range Alaska to California.

Edible Uses The leaves of this widespread plant are used to flavor meats and jellies. Field mint can be made into a soothing tea, cooked as greens to be eaten alone or added to soups and stews as a flavoring.

To make sipping teas, plants were gathered from shaded locations, where they have fewer stems and produce a better-tasting tea. To make tea, simply place a small handful of the dried leaves in a teapot with boiling water and steep for a few minutes.

Medicinal Uses Field mint tea infusion is thought to be quite good for digestion. Mint tea was used by Native peoples to treat colds, coughs, fevers, stomach pains, vomiting, kidney problems and headaches. Menthol is an essential oil of mint that brings some relief from pain.

Hot mint-leaf poultices were also used to relieve arthritis and rheumatoid arthritis. Leaves were also packed around aching teeth. For medicinal use, plants were gathered from above the low watermark in areas exposed to the sun, where they have a higher essential oil content. The best time to gather this plant is in late summer.

Precautions Do not drink large amounts of mint tea during pregnancy or heavy menstruation.

Notes Field mint is an aromatic plant that was hung in dwellings as an air-freshener, and also used as a perfume for the body. Powdered mint leaves were sometimes sprinkled on berries and drying meat to repel insects.

Self-heal *Prunella vulgaris*

Other Names Also known as all-heal, blue curls, brownwort, brunella, carpenter's herb, carpenter-weed, common self-heal, common woundwort, dragonhead, heal-all, hook-heal, hook weed, self heal, sickleweed, sicklewort, slough-heal, heart of the earth.

Mint Family (Lamiaceae)

Description Overall: Perennial herb. **Flowers:** Purplish-blue to pink or white, short-stalked; corolla has an upper lip and a lower lip with 3 parts, to 0.6" (15 mm) long, spike-like cluster on top stem. **Fruits:** 4 nutlets. **Leaves:** Ovate to lanceolate. **Other:** Plants grow from rhizomes or creeping stem base.

Size To 19.7" (50 cm) high but often much smaller.

Habitat In moist areas including clearings, fields and forest edges; common at low to mid-elevations.

Range A ubiquitous worldwide species.

Edible Uses Fresh leaves can be cut up and steeped in cold water to make a refreshing drink. They can also be dried, ground and stored as a powder to make cold drinks.

Medicinal Uses This plant is commonly used to heal cuts, bruises and skin inflammations simply by applying the juice to the affected area. It closes wounds effectively, promoting healing. The juice is also used for boils. The entire plant was boiled by Nuxalk peoples and taken as a weak tea for the heart.

The flowers and leaves are harvested when in full bloom. Both have been used in making a poultice for cuts, bruises and skin inflammations. The leaves and flowers have been infused in water and gargled for sore throat, as well as throat irritations and hemorrhages. It is also sweetened with honey, and used to treat a sore throat or mouth.

Notes Self-heal is a mixture of native subspecies (ssp. *lanceolata*) as well as those introduced from Eurasia (ssp. *vulgaris*). This plant has been used by the Quinault, Quileute and other aboriginal groups around the world to treat a wide variety of conditions.

Common Yarrow *Achillea millefolium*

Other Names Also known as blood-wort, cammock, carpenter's grass, devil's plaything, dog daisy, gordoloba, green arrow, knight's milfoil, milfoil, milfoil thousand-leaf, millefoil, nosebleed, old man's pepper sanguinary, soldier's woundwort, soldier's woundwort, thousand-leaf, thousand-leaved clover, thousand-seal, yarrow; includes *A. borealis, A. lanulosa* (the species native to western North America).

Description Overall: Aromatic perennial herb. **Flowers:** Cream-colored to pink ray flowers, to 0.2" (5 mm) across; yellow disc flowers, clustered. **Leaves:** Fern-like, feathered, alternate, dissected 2–3 times.

Size To 32" (80 cm) high, normally much smaller.

Habitat Dry to moist, well-drained, open areas including meadows, rocky slopes, gravel bars, roadsides and clearings.

Range Alaska to New Mexico.

Medicinal Uses Yarrow tea and poultice have been used externally to treat many maladies including burns, boils, sores, pimples, earaches, sore eyes and mosquito bites. Yarrow tea has also been taken internally for colds, diarrhea and fevers. Yarrow was best known, however, as a plant that stops bleeding. In fact, it has saved the lives of many soldiers by being applied to their wounds. An anti-inflammatory can be prepared by steeping the herb in water and adding the mixture to bathwater, which has been used to relieve rheumatoid arthritis and other inflammations of the joints. This species is considered an astringent, alterative, diuretic and tonic.

Yarrow poultices are useful for varicose veins, especially during pregnancy. Various Native groups have used medicinal preparations of yarrow for sore throat, colds and coughs, bronchitis, childbirth, diarrhea and other troubles. The entire plant is best harvested in late summer when in full bloom, and allowed to air-dry completely in large pieces. It should not be stored in plastic bags.

Precautions Individuals with sensitive skin may react to this species. Large quantities can increase blood supply, so it is not advisable for extended use during pregnancy. Yarrow contains thujone, a toxic substance, and large doses can cause a miscarriage.

Notes This amazing herb has been used to treat an extended list of ailments since the 15th century, much as aspirin and acetaminophen are used today. Yarrow is also well known as a fumigant and insecticide.

Similar Species Siberian Yarrow *A. sibirica* has leaves with sharp teeth that are not dissected. Present at low elevations.

Pineapple Weed *Matricaria discoidea*

Other Names Also known as false chamomile; formerly classified as *M. matricarioides*.

Aster Family (Asteraceae)

Description Overall: Annual herb, pineapple-scented. **Flowers:** Lacks rays; disc florets yellowish, to 0.4" (1 cm) across, cone-shaped with several flower heads in branching clusters; flowers throughout summer. **Fruits:** Achenes; pappus is a short membranous crown. **Leaves:** Alternate, freely branching, each divided into several fine, linear segments, each to 2" (5 cm) long. **Other:** Root system fragile.

Size To 16" (40 cm) high.

Habitat Disturbed ground, waste areas, roadsides and similar sites; plains to montane.

Range Southern Yukon and NWT to New Mexico.

Edible Uses Pineapple weed is a tasty snack, as a finger food or in salads.

Medicinal Uses Pineapple weed tea has been used in the treatment of colds, upset stomachs, fevers, diarrhea and menstrual cramps. The flowers and leaves are often collected and dried, mainly to make tea, but the entire plant can be utilized.

Women have taken this tea at childbirth and to aid in delivering the placenta. The tea has also been used for stomach aches, as a mild relaxant and for colds, flatulence and menstrual problems. The juice of a fresh flower has been squeezed directly into the eye to treat eye infections.

Precautions Some people are allergic to this species, as well as other members of the Aster Family.

Notes As its name suggests, this species has a noticeable pineapple scent. Pineapple weed is native to western North America, and has spread as a weed to the east and to Eurasia. It has been used in an amazing range of ways: as bait in a lynx trap when the leaves are mixed with other materials, as an insect repellent, as a means of retarding spoilage and repelling insects when ground up and sprinkled on meat or berries. The Blackfoot people used dried pineapple weed as a perfume. The plant contains essential oil, up to 0.45% present in the leaves and nearly double that in the flowers. This oil consists primarily of beta-farnesene, geranyl isovalerate, germacrene and myrene.

Similar Species Chamomile *M. chamomile* is a close relative, with white rays. It is uncommon east of the Cascades.

Ox-eye Daisy *Leucanthemum vulgare*

Other Names Also known as great ox-eye, goldens, marguerite, moon daisy, horse gowan, maudlin daisy, field daisy, dun daisy, butter daisy, horse daisy, maudlin-wort, white weed; formerly classified as *Chrysanthemum leucanthemum*. **Aster Family (Asteraceae)**

Description Overall: Perennial herb. **Flowers:** Flower head solitary at the tips of branches, to 2" (5 cm) across, ray flowers white, discs yellow. **Fruits:** Achenes black, with approximately 10 white ribs. **Leaves:** Alternate, somewhat hairy, spoon-shaped, toothed to incised. **Other:** Root is short and somewhat creeping.

Size To 28" (70 cm) high.

Habitat Present in roadsides, vacant lots and similar sites, also in fields, meadows and mountains.

Range Alaska to New Mexico.

Edible Uses The young leaves and stem leaves are edible and very sweet, a great addition to salads. Wine can be made from the flower heads.

Medicinal Uses Ox-eye daisy has been effective in treating whooping cough, asthma and nervous excitability. The leaves, stalks and flowers can be boiled with honey to produce a drink for the treatment of chronic cough and bronchial inflammation. Taken internally, ox-eye daisy has been used in a variety of ways for treating night sweats, jaundice, mouth and vocal cord swelling, liver and gallbladder problems, loss of appetite, spasms, edema and skin swelling.

Externally ox-eye daisy has been used to treat ulcers, cuts, bruises and conjunctivitis. The best time to collect this species is when it is first flowering, but it can be harvested all summer. The entire plant or just the fresh flowers and stems can be collected. The herb can be thoroughly dried and stored up to a year.

Precautions Some people have an allergic reaction to ox-eye daisy and other members of the Aster family. The plant may also irritate the skin.

Notes Ox-eye daisy is a species introduced from Europe. Its foliage emits a scent similar to yarrow or chamomile when crushed. The most effective medicinal part of this plant is the flower. Ox-eye daisy has an essential oil containing chrysanthenone, verbenone, pyrethrins and over twenty known polyacetylenes.

Common Dandelion *Taraxacum officinale*

Other Names
Also known as blowball canker-wort, chicoria, conseuelda, doonheadclock, fortuneteller, Irish daisy, lion's tooth, pissabed, priest's crown, puffball, swine snort.

Aster Family (Asteraceae)

Description
Overall: Perennial herb. **Flowers:** Bright yellow with ray florets only, to 2" (5 cm) across. **Fruits:** Achenes in a cottony tuft replace the flower head. **Leaves:** Basal, oblong, lobed.

Size To 24" (60 cm) high.

Habitat On lawns and waysides.

Range Now global in its distribution; believed to have originated in Greece.

Edible Uses Tender young leaves are commonly used as salad greens, or cooked as a vegetable. The brown taproot is preferably gathered in late summer or fall. It can be cooked and eaten as a vegetable, or cleaned thoroughly, dried in the oven at a low temperature and ground into a caffeine-free coffee substitute. The flowers can be made into fritters but are more often used to make wine. The entire plant has also been used to make beer.

Medicinal Uses Dandelion greens, a healthy addition to the diet and sometimes called the poor man's ginseng, are a rich source of beta carotene, which is believed to be a potent protector against cancer. The plant is believed to aid in curing liver disease and kidney stones, and to aid in lowering high blood pressure, easing bowel functioning, relieving anemia and controlling diabetes. According to one authority, Michael Moore, a "fresh root tincture is good as an anti-inflammatory for hives, arthritis, and other extended allergic reactions. The root stimulates bile, and is both a laxative and a liver tonic, but without the irritation induced by most bile stimulants."

Precautions Be careful in choosing sites to harvest dandelion plants. Many lawns and waysides are sprayed with herbicides to control weeds. This plant's milky sap has been known to cause rashes on very sensitive skin.

Notes Dried and powdered dandelion root makes a good coffee substitute by itself, or mixed with chicory root, mint leaves, green pine needles and/or yarrow leaves. The common dandelion is not only edible, it is also a valuable medicinal plant, with so many uses that an entire book has been written about it: *The Dandelion Celebration* by Peter Gail. You may think about dandelions differently the next time you notice them in your lawn!

Chicory *Cichorium intybus*

Other Names
Also known as blowball, blue dandelion, blue daisy, blue sailors, French dandelion, wild endive, wild bachelor button, ragged sailors, succory, wild succory, witloof.
Aster Family (Asteraceae)

Description
Overall: Perennial herb.
Flowers: Blue ray florets only, approximately 1.6" (4 cm)
wide, stalkless clusters of 1–3 on branches. **Fruits:** Hairless, 5-sided achenes, to 0.1" (3 mm) long. **Leaves:** Basal, dandelion-like, lanceolate and deeply toothed, 8" (20 cm) long. **Other:** Plants arise from deep, fleshy taproots that are often thick and forked.

Size To 4.9' (1.5 m) high.

Habitat Roadsides and disturbed ground, fields and waste areas; plains to mid-elevation.

Range Southern BC to California and New Mexico.

Edible Uses Young roots can be eaten raw. Young or mature roots can be roasted, boiled, steamed or added to stews. The roots are often dried in the sun or oven, then roasted and ground to make a coffee substitute, or mixed with ground coffee. Fresh chicory leaves are a great addition to salads, or they can be cooked as a potherb. They are very good blanched. Gather fresh young leaves in spring. The roots are best collected before flowering begins, because they grow increasingly bitter as the season progresses.

Medicinal Uses Herbalists recognize chicory as a diuretic, laxative and hepatic. It has been used as an appetite stimulant, cholagogue, diuretic, tonic, hepatic and (with water) laxative. It has also been used for to treat jaundice and spleen problems. The leaves were crushed and made into poultices for inflammations, irritations, rashes, swellings and even for inflamed eyes.

Precautions Extended or excessive use of this species may slow down digestion, damage the retinas, cause a rash or bring on side effects in nursing women. Consult a doctor before using this plant medicinally, especially with individuals having gallstones or other gallbladder diseases.

Notes Chicory is an introduced plant, native to Europe but widespread today. An analysis of 100 grams of fresh chicory greens showed that it contains 86 mg of calcium, 40 mg of phosphorus, 420 mg of potassium, 22 mg of vitamin C, and a healthy 4,000 IU of vitamin A.

Canada Goldenrod *Solidago Canadensis*

Other Name Also known as meadow goldenrod.

Aster Family (Asteraceae)

Description Overall: Perennial herb. **Flowers:** Yellow, diamond-shaped mass. **Fruits:** White, short-haired achenes. **Leaves:** Lance-linear, 3 veins on each partly toothed leaf. **Other:** Plants arise from long, creeping rhizomes.

Size To 60" (150 cm) high.

Habitat Roadsides, disturbed sites and forest openings.

Range Alaska to California.

Medicinal Uses An infusion can be made using one teaspoonful of the flowers in a cup of water to relieve symptoms of hay fever and other allergies. Kidney, bladder problems and constipation can be treated by boiling the leaves and stems into a decoction. This same substance can be cooled and used as a wash to help dry out weeping sores. Diarrhea has been treated using the flower heads in a tea. Be sure to gather the flowers in the bud stage rather than in the flowering stage.

Precautions Allergies associated with all goldenrods are well known. A healthcare professional should be consulted before using goldenrod if urinary tract disorders are present. Goldenrod should not be used for those that are retaining water with heart or kidney disorders.

Notes It seems ironic that one of the best-known plants that cause allergies can be a remedy for the same and related problems, but it has been shown to be true for some people. There are numerous species of goldenrods in North America. Canada goldenrod is the most common species in the Northwest.

Palmate Coltsfoot *Petasites palmatus*

Other Names Also known as coltsfoot, butterbur, palm-leaved coltsfoot; formerly classified as *Petasites frigidus*, var. *palmatus*, *Tussilago palmatum*.
Aster Family (Asteraceae)
Description **Overall:** Perennial herb. **Flowers:** Cream-colored, clustered. **Fruits:** White, hairless achenes with 5–10 ribs. **Leaves:** Basal, palmate, deeply divided with 4–7 lobes. **Other:** Plants arise from rhizomes.
Size To 20" (50 cm) high.
Habitat Wet meadows, moist roadsides.
Range Alaska to California.
Edible Uses The young flowering stems and leaves can be steamed, or stir-fried as a vegetable. The leaves are best gathered from mid-June to late August, and the stems should be cut at ground level. Palmate coltsfoot was also considered to be an important salt substitute. The Siberian Eskimo are known to have roasted and eaten the rootstocks.

Medicinal Uses A root poultice was applied to sprains, contusions and similar troubles. The roots were dried, coarsely grated and applied to boils and rubbing sores to help dry them up. The roots have been used to make medicinal teas in the treatment of tuberculosis, asthma, rheumatoid arthritis, sore throat and stomach ulcers.
Precautions Coltsfoot should not be consumed in large quantities as it contains alkaloids that can be harmful. Pregnant women should not consume this species.
Notes Palmate coltsfoot is one of the first herbs to appear in the spring. Its delightful flowering stalk is the first part of the plant to appear!

Pearly Everlasting *Anaphalis margaritacea*

Other Names
Also known as everlasting, life everlasting; formerly classified as *Gnaphalium margaritacea, Antennaria margaritacea.*
Aster Family (Asteraceae)
Description **Overall:** Perennial herb.
Flowers: Yellow center surrounded by whitish bracts, forming a flat-topped cluster.
Fruits: Very small, roughened achenes; short white pappus hairs. **Leaves:** Alternate, lance-shaped, conspicuous midvein, green above, white-woolly beneath.
Other: Plants arise from rhizomes and off of parallel woody underground roots.
Size Normally to 24" (60 cm) high.
Habitat Dry, open sites, rocky slopes, clearings, meadows, fields; from low to subalpine.
Range Alaska to California.
Edible Uses The leaves and young plants make a useful potherb.
Medicinal Uses The roots of this plant and five or six small shoots were steeped in hot water as a tea for digestive ailments. The tea caused vomiting as well as having a laxative effect to clean out the stomach. Medicinal teas made from pearly everlasting have also been used to treat coughs, colds, diarrhea, dysentery, throat infections and upset stomachs. Michael Moore states that the leaf poultice and, even better, the flower poultice have useful properties in "healing sunburns and other moderate burns from heat and friction." Poultices have been used to treat a number of other problems including bruises, burns, rheumatoid arthritis, sores and swellings. The leaves and on occasion the whole plant have been made into a poultice that is said to soothe and relieve the redness and swelling of bruises and contusions.

Notes Pearly everlasting is a common species, which was also smoked like tobacco by Native peoples. Its beautiful flowers are always a pleasure to see.

Stinging Nettle *Urtica dioica*

Other Names Also known as nettle; includes *U. lyallii* and *U. gracilis*.
Nettle Family (Urticaceae)
Description **Overall:** Perennial herb. **Flowers:** Greenish to whitish, tiny in dense clusters that droop from the stalk, male and female flowers in separate clusters on same plant. **Fruits:** Flat, lens-shaped achenes, 0.06" (1–2 mm) long. **Leaves:** Opposite, lance-shaped to oval, coarsely saw-toothed; 0.2–0.6" (5–15 mm) long. **Other:** Plants arise from rhizomes.
Size To 10' (3 m) high.
Habitat Moist meadows, open forest, disturbed sites; lowlands to subalpine.

Range Alaska to California.

Edible Uses Stinging nettle is sometimes called "Indian spinach." The young leaves are often cooked and eaten as greens, and the tender young shoots can be eaten like spinach and are considered delicious. The young plants are also used to make nettle tea, and beer and wine that have earned the rating "good." The roots, best gathered in fall to spring, can be roasted. Stinging nettle is rated high in vitamins A, C and D.

Cooking or drying destroys the stinging properties of nettles. Be careful, however, of how much you eat—large quantities may cause a mild burning sensation.

Medicinal Uses Stinging nettle is often used to treat internal bleeding, especially for those who have bleeding of the lungs and stomach.

Precautions Always use caution in the collection of stinging nettle. Gloves, long sleeves and long pants must be worn. Stinging nettle should not be used by diabetics or pregnant women.

Notes The stiff, hollow hairs along the leaves and stems of this plant contain formic acid, a strong irritant to the skin. When any of these hairs are broken, the acid is injected into the skin, causing a rash, itching and blisters that may last for a few minutes or as long as a couple of days

Lamb's Quarters *Chenopodium album*

Other Names Also known as baconweed, Belgian spinach, fat hen, goosefoot, middens miles, muchweed, pigweed, smooth pigweed, white chenopodium, white goosefoot, wild spinach, wormseed.

Goosefoot Family (Chenopodiaceae)

Description **Overall:** Annual herb. **Flowers:** Pale green, tiny; in dense clusters or spikes, mealy, in the leafaxils and at the stem tips. **Fruits:** Black, shiny lens-shaped seeds, to 0.06" (1.6 mm) long, single seed enclosed within a thin, paper-like envelope. **Leaves:** Alternate, somewhat fleshy, lance- to diamond-shaped, to 4" (10 cm) long, grayish green and very mealy below, irregularly toothed or lobed. **Other:** Plants grow from slender taproots.

Size Normally to 3.3' (1 m), occasionally to 6.6' (2 m) high.

Habitat Disturbed sites including cultivated and waste lands, gardens, fields, roadsides; low to mid-elevation.

Range Southwestern BC to California.

Edible Uses The young leaves, often said to taste like spinach, are high in vitamin C and make an excellent vegetable dish, but they are also a mild laxative. The leaves and tender tips can be steamed or cooked with onions or garlic to prepare a wonderful hearty dish.

The seeds can also be harvested if the outer bracts are separated from the seeds. Traditionally this was done by lightly grinding the fruits, which loosened the outer bracts. The seeds were then washed, dried and ground into flour.

Medicinal Uses A tea was made of the leaves or the entire plant by Native peoples to relieve stomach pains. A decoction was taken internally or applied externally to treat painful limbs.

Strawberry Blite *Chenopodium capitatum*

Other Names Also known as blite, Indian paint, Indian strawberry, strawberry-blite, strawberry spinach. **Goosefoot Family (Chenopodiaceae) Description** Overall: Annual herb. **Flowers:** Deep red, pulpy, tiny, with 3–5 fleshy sepals and no petals; flower clusters in round heads to 0.6" (1.5 cm) wide. **Fruits:** Black seeds, enclosed in a red, fleshy, berry-like calyx. **Leaves:** Alternate, dark green, arrowhead-shaped, to 3.9" (10 cm) long, toothless or wavy-toothed. **Other:** Plants arise from slender taproots. **Size** To 19.7" (50 cm) high. **Habitat** A variety of sites including dry and well-drained to moist ground, burned damp clearings and similar sites; foothills to subalpine.

Range Yukon to New Mexico.

Edible Uses The bright red, fleshy, sweet flower clusters are edible raw and are rich in calcium, protein and vitamins, but they should be eaten in moderation. The fruits can also be cooked. Young plants can be cooked as a potherb or added to stews and soups.

Medicinal Uses Strawberry blite has been used externally to treat throat and mouth ulcers. It has also been injected as a treatment for leucorrhea and hemorrhages of the bowels.

Precautions This species can be eaten raw, but large quantities can be toxic! Moderation is the key with strawberry blite, as with most wild foods. The toxins are rendered harmless once the plant is cooked.

Notes Strawberry blite is a distinctive plant, rich in vitamins A, BI, B2, B6 and C, as well as a fine source of calcium, iron and potassium. Its distinctive bright red flowers make it very knoticable from a distance.

Bunchberry *Cornus canadensis*

Other Names Also known as Canada dogwood, dwarf cornel, dwarf dogwood, pigeonberry; formerly classified as *C. unalaschkensis*. **Dogwood Family (Cornaceae)**
Description **Overall:** Perennial herb. **Flowers:** Greenish white to purple cluster of flowers, surrounded by 4 greenish white bracts; to 2" (5 cm) diameter; May and June. **Fruits:** Red berry-like drupes; to 0.3" (8 mm) diameter; July and August. **Leaves:** Terminal whorl, oval; to 3" (7.5 cm) long. **Other:** Plants arise from rhizomes.
Size To 10" (25 cm) high.
Habitat Moist woods and clearings.
Range Alaska to Greenland, south to California; Asia.
Edible Uses The fruit is edible, although it is bland. Native peoples sometimes added the berry-like drupes of other fruit to improve the taste.
Medicinal Uses Bunchberry has been used to decrease fever, inflammation, and pain, as aspirin does but without the stomach irritation or allergic reactions sometimes caused by balsam poplar, aspen, willow, wintergreen and other plants. Some Native peoples sprinkled the ashes of bunchberry leaves on sores. This plant has also been reported to reduce the effects of poisons. Bunchberry is considered a mild and predictable herb for colitis, chronic gastritis, diarrhea, dysentery and other conditions.
The entire plant was traditionally gathered for drying from summer to early fall.
Precautions Be sure to eat only ripe berries. Unripened berries will likely cause stomach aches.
Notes One of the many other uses of bunchberry is to chew the berries for treatment of "insanity." It is often picked and eaten as a trailside snack while hiking.

Common Plantain *Plantago major*

Other Names Also known as broad-leaved plantain, whiteman's foot.

Plantain Family (Plantaginaceae)

Description **Overall:** Perennial herb. **Flowers:** Petals tiny and greenish, in long spikes on tall stalks to 12" (30 cm) long, dense spike at top of flowering stalk. **Fruits:** Many-seeded capsules, top splits off like lid. **Leaves:** All basal; oval blades with prominent parallel veins, to 6" (15 cm) long; somewhat fleshy. **Other:** Plants grow from short, thick rhizomes.

Size To 20" (50 cm) high.

Habitat On disturbed ground and in waste lands; low to mid-elevation.

Range Across most of North America.

Edible Uses The seeds have been ground into meal or flour and used to make bread or pancakes, but this is thought not to be a widespread practice.

Medicinal Uses The crushed leaves of common plantain have been placed on wasp stings, causing the pain to disappear in less than 10 minutes. This treatment has also worked for bee stings and the bites of other insects, spiders and poisonous reptiles. Numerous Native peoples have used common plantain to make a poultice that soothes cuts, sores and burns.

The Cree chewed the leaves of common plantain to relieve toothache. They also ate the leaves to stop internal bleeding. This species has also been used to treat gastric ulcers. The fresh juice from its leaves was said to relieve stomach problems and to expel worms. Kidney, bladder, gastrointestinal and respiratory problems are treated with a medicinal tea. This species is considered an alterative, astringent, diuretic and antiseptic.

Notes Common plantain is an introduced weed that has a remarkable history for the many ailments that it has been used to treat. Preliminary studies conducted on this species indicate that plantain seeds may lower blood pressure and blood cholesterol levels.

Western Skunk Cabbage *Lysichiton americanus*

Other Names Also known as swamp cabbage, swamp lantern, yellow arum, yellow skunk cabbage; also classified as *Lysichiton americanum*.
Arum Family (Araceae)
Description Overall: Perennial herb. **Flowers:** Greenish yellow; on a spike with a single hood-like bract. **Fruits:** Berry-like, greenish, embedded in and difficult to separate from the fleshy flower spike. **Leaves:** Basal, in a large rosette, oblong. **Other:** Plants arise from fleshy rootstocks.
Size To 5' (1.5 m) high.
Habitat Wet areas including swamps, fens, muskeg, wet forest, wet meadows; low to mid-elevation.
Range Along the coast from Alaska to central California.
Edible Uses The young spring leaves make a good potherb if gathered just as they arise fresh from the earth. The whole roots were harvested by Native peoples and roasted in pits. The unpleasant qualities of this plant were greatly reduced by the heat and moisture. The roots were then dried and ground into a flour as an emergency ration. The peppery taste was reduced further by storing the flour a week or more before using it.

Medicinal Uses Western skunk cabbage has been used to help alleviate spasmodic and painful cramps, bouts of coughing and gagging, and asthma occurring under stressful conditions. To make this species palatable, the roots were chopped and simmered with 4 parts honey for a couple of hours. The mixture was then cooled, strained and bottled.

Precautions Nausea, gastric irritation and diarrhea often result from consuming too much western skunk cabbage.

Round-leaved Sundew *Drosera rotundifolia*

Sticky-tipped leaves.

Other Name Also known as sundew.
Sundew Family (Droseraceae)
Description Overall: Insectivorous perennial herb. **Flowers:** White, only opens in full sunlight; to 0.1" (4 mm) across; July. **Fruits:** Many-seeded capsules. **Leaves:** Glandular in a basal rosette, to 2.8" (7 cm) long, with reddish glandular hairs (tentacles) that exude drops of sticky fluid. **Other:** Plants arise from a shallow root system.
Size To 10" (25 cm) high.
Habitat Sphagnum bogs, fens, wet meadows; low to mid-elevation.
Range Alaska to California.
Medicinal Uses An antibiotic agent is present in the sap of round-leaved sundew that was effectively used in treating patients for several bacteria. Tuberculosis, asthma, bronchitis and coughs have also been treated using round-leaved sundew. The leaves were used to treat corns, warts and bunions by several Native groups of the Northwest. The irritating fluid produced by this plant, which contains an enzyme and the antibiotic plumbagin, has been used for centuries in the removal of warts and corns. This species is considered a stimulant, expectorant, demulcent and antispasmodic.

A fresh herb tincture has been used for coughing bouts. Whooping cough has also been treated extensively with this species in homeopathic medicine.
Precautions External use of this plant has been known to cause water blisters.
Notes The round-leaved sundew captures insects with sticky-tipped hairs and digests the insect's nutrients, a process that takes 24 to 48 hours. This species is uncommon in the southern part of its range, so restraint should be exercised in removing plants for medicinal use.

Common Cattail *Typha latifolia*

Other Names Also known as cattail, reedmace.

Description Overall: Perennial herb. **Flowers:** Tiny, forming dense, cylindrical spikes in separate male and female clusters, male above female spike. Both lack petals and sepals. Male flower forms a short-lived yellowish terminal spike, to 6" (15 cm) long. Female flowers grow in a green to brown cluster. The fruiting body replaces the female cluster by late summer. **Fruits:** By August, the brown female spike contains thousands of tiny fruits covered with white hair. **Leaves:** Alternate, grass-like, to 0.8" (2 cm) wide, unbranched. **Other:** Plants arise from coarse rhizomes. Plants spread rapidly along wetlands under good conditions.

Size To 9.8' (3 m) high.

Habitat Wetlands of various types; still water and slow-moving water.

Range Southern Yukon and NWT to California.

Edible Uses Common cattail is one of the more important food plants in all of North America. Pollen, collected from male flower spikes, can be dried, ground and used to extend flour in baking pancakes, muffins, breads and other items. Cattail pollen alone is not sticky enough to work in baking, so it must be mixed with equal parts wheat flour. Collecting pollen is best done by harvesting the upper male flower spikes (found above the female spikes) early in the season. Then take them home and shake the pollen from them repeatedly each day for several days, carefully drying it as it is collected. When it is well dried, the pollen stores well for up to one year.

Young female cattail flower spikes can be cooked and eaten like corn on the cob. They even taste like corn on the cob! The tender inner core of young shoots can also be eaten, raw or cooked. The white center of the roots can be eaten raw, roasted or boiled, and it can be dried and ground to make a flour.

Medicinal Uses Cattail has been used medicinally by several aboriginal groups, including the Blackfoot people, who used cattail fluff as an antiseptic to treat burns and scalds. The Cree people cut up and ate the stem to relieve diarrhea.

Precautions Western blue flag *Iris missouriensis* also grows in similar environments, and its roots are poisonous. They should never be eaten. This plant has a distinctive blue flower. Be sure of the identity of any plant whose roots you are harvesting.

Notes In a survival situation, it is useful to know that the down from mature cattail seeds makes a perfect tinder. One cattail stalk is estimated to produce up to 220,000 seeds in one season.

Similar Species Narrow-leaved Cattail *T. anigustijolia* produces narrower leaves, to about 0.2" (5 mm) wide. It is edible, but not as desirable to eat.

Yellow Pond Lily *Nuphar lutea*

Other Names
Also known as yellow water lily, cow lily, spatterdock, Indian pond lily; formerly classified as *N. luteum*, var. *polysepalum*, *Nymphaea polysepala*.
Water-lily Family (Nymphaeaceae)

Description
Overall: Aquatic perennial. **Flowers:** Yellow, cup-shaped, waxy; to 1" (2.5 cm) wide. **Fruits:**

Egg-shaped capsules, ribbed, to 3.5" (9 cm) long, leathery then decaying to release seeds. **Leaves:** Heart-shaped, waxy green leaves with long stems. **Other:** Plants arise from extensive rhizomes.

Size To 79" (2 m) high.

Habitat Ponds, small lakes and similar quiet waters.

Range Alaska to Colorado.

Edible Uses The ripe seeds were often extracted from the pods and roasted, or sometimes ground into flour or meal. They can be popped like corn in hot oil and are very nutritious and palatable. Aboriginal peoples included the seeds in their diets. Thin rhizome slices were also dried and ground into meal for making a gruel or for thickening soups.

Considerable differences have been published regarding the edibility and taste of yellow pond lily rhizomes. Some accounts deem them unsuitable due to their bioactive alkaloid content. Others say that this species is edible but bitter, and still others describe it as sweet-tasting. Thin rhizome slices were also dried and ground into meal for making a type of gruel or for thickening soups.

Medicinal Uses Native peoples in what is now Montana used the rhizomes to treat sexually transmitted disease. The rhizomes were also boiled and the liquid drunk, and a crushed rhizome poultice was applied to affected areas. Gallstones were treated with the yellow pond lily by the Yukon Tagish. Other aboriginal groups used the rhizomes to combat tuberculosis and other diseases

Precautions This plant was not often used because it was difficult to gather at most locations. Because the possible bioactive alkaloid content is a concern, do not eat excessive quantities of the rhizomes until it is clear how edible this plant is. **Caution is advised!**

Notes It is much easier to harvest the seeds of this common plant than the rhizomes. It is also much less disruptive to the plant and area.

Horsetail *Equisetum* spp.

Scouring rush, *Equisetum hyemale*.

Horsetail Family (Equisetaceae)

Description Overall: Perennial herb with jointed, hollow stems. **Leaves:** Appear as small scales, 8–12 fused around stems. **Other:** Grows from spreading rhizomes with tubers.

Size To 4.9' (1.5 m) high.

Habitat Moist to wet open sites, disturbed ground and on occasion dry, wooded sites; lowlands to alpine.

Range Alaska to New Mexico.

Edible Uses The young heads of horsetails can be boiled for about 20 minutes with a change of water halfway through. Do not consume large quantities — horsetail can have a toxic effect (see below). Once boiled, it can be dipped in an egg and crumb mixture, then fried. Horsetail tubers are also edible and can be eaten raw in the early spring, or boiled if collected later in the season.

Medicinal Uses A poultice of horsetail was often applied to help wounds heal. Horsetail is used today to treat eye and skin problems, because of its high silica content. Equisetum silica tablets are taken for catarrhal conditions such as pus-like discharges from the ear, nose and throat. It has been used for other disorders, including offensive perspiration of the feet.

Horsetail has a reputation for easing the discomfort of difficulty in urination, and all types of internal bleeding. Internal symptoms have been treated with a horsetail tea boiled for 45 minutes and allowed to cool. The tea was also used externally to treat sores, skin problems, wounds and inflammation of mouth and gum tissue. The entire plant above ground was harvested fresh for medicinal purposes.

Precautions There is much controversy about the usefulness of horsetail species and the potential dangers associated with their use. Symptoms of poisoning have been noted with excessive consumption (over 8 oz/225 g). Several chemicals in horsetail have a slightly toxic effect—normally the destruction of thiamine (a B vitamin). If you feel you have consumed too much *Equisetum*, take vitamin B to reverse any side effects.

Notes Several species of horsetail are found in this region. They do not have flowers, but cones grace the tips of their stalks where the spores are produced. The stems of horsetails are embedded with silica crystals, which makes them very abrasive. They were used by Native peoples to polish pipes, bows and arrows. Horsetails have caused livestock deaths, but many wild animals, including bears, caribou, moose and sheep, eat these plants regularly.

Western Bracken Fern *Pteridium aquilinum*

Other Names Also known as brake, brake fern, bracken, eagle fern, hog-brake, pasture-brake, western bracken, western brake-fern.

Bracken Family (Dennstaedtiaceae)

Description Overall: Deciduous fern. **Sori:** Marginal, continuous, covered by rolled leaf edge. **Leaves:** Fronds large, solitary, triangular blades, 2–3 times divided pinnate, leaflets 10+ pairs, normally opposite. **Other:** Plants arise from rhizomes.

Size To 10' (3 m) high.

Habitat Meadows, roadsides, clearings, burns, avalanche tracks, dry to wet forests; low to subalpine elevation.

Range Alaska to California.

Edible Uses This species has been used as a vegetable, but this is now discouraged due to its apparently harmful effects (see below).

Medicinal Uses Due to the warnings issued about this species, western bracken fern should not be taken internally. In the past it was used internally to expel intestinal worms, increase urine flow and relieve stomach cramps.

   The leaves of this fern have been used externally to speed the recovery of broken bones, and to make a steam bath to relieve symptoms of arthritis. The entire plant was harvested at maturity.

Precautions Native peoples formerly ate this fern, but consumption is now strongly discouraged. The plant has been found to contain a carcinogenic substance that can cause stomach cancer. Cooking may remove this substance, but further research is required. Therefore a **WARNING** remains in effect. The stems, rhizomes and foliage are all thought to be harmful. Western bracken fern is also implicated in livestock poisoning.

Notes There are edible ferns in the Northwest, but this is NOT one of them. Wild fiddle-heads ONLY from the ostrich fern can safely be eaten (see p. 72).

Western Sword Fern *Polystichum munitum*

Other Name Also known as sword fern.

Wood Fern Family (Dryopteridaceae)

Description Overall: Evergreen fern. **Sori:** Large, circular, located midway between the margin and midvein. **Leaves:** Blade lance-like, once-pinnate; leaflets alternate, sharp-toothed. **Other:** Plants arise from scaly rhizomes.

Size To 5' (1.5 m) high.

Habitat Moist forests; low to mid-elevation.

Range Southeast Alaska to California.

Edible Uses The rhizomes were once steamed in a traditional pit oven, or boiled or baked in coals, then eaten. This was often considered a starvation food.

Medicinal Uses The raw plant was chewed and eaten to treat sore throat or tonsillitis. Women chewed the leaves to help with childbirth. A poultice of chewed leaves was sometimes applied to sores and boils. The spore sacs were made into a poultice and applied to burns. Sores were washed with an infusion of stems. Cancer of the womb was treated by chewing on the young shoots of this fern.

Notes Western sword fern is a stout species whose fronds were used to line earth ovens, and to make mattresses.

Similar Species Ostrich Fern *Matteuccia struthiopteris* is a similar-looking species with deciduous leaves and sori that are present on brown pod-like leaflets found on separate fertile leaves. The fiddleheads are springtime favorites for those who dine on our indigenous plants.

Licorice Fern *Polypodium glycyrrhiza*

Other Names Also known as common polypody, rock brake, sweet fern; formerly classified as *P. vulgare*, var. *occidentale*, *P. occidentale*.

Polypody Family (Polypodiaceae)

Description Overall: Evergreen fern. **Sori:** Oval to round, 1 row on each side of the main stem; not enclosed or covered by an indusium (membrane). **Leaves:** Stipes normally shorter than the fronds; once-pinnate leaflets, fronds to 12" (30 cm) long and 3" (7.5 cm) wide. **Other:** Plant arises from a creeping rhizome.

Size To 28" (70 cm) high.

Habitat Normally epiphytic on deciduous tree trunks, especially bigleaf maple, but also on wet, mossy ground, nurse logs; primarily at low elevation.

Range Alaska to Oregon.

Edible Uses Licorice fern rhizomes were gathered and chewed for their flavor by several Native groups. They were eaten raw, or dried, steamed, scorched or otherwise prepared. This fern was also used as a sweetener for bitter-tasting medicines. Rhizomes were gathered in late summer to mid-fall.

Medicinal Uses The rhizomes of licorice fern were an important medicine for treating colds, cough, sore throat and respiratory ailments such as tuberculosis. To treat a cough, a piece of the raw rhizome was simply chewed and the juice was swallowed. This species has also been used to treat stomach troubles.

Precautions Licorice fern likely contains salicylates, and therefore is best avoided by those who have aspirin allergies or blood dyscrasias, or who take anticoagulant medications. It also should not be used during pregnancy.

Notes The rhizome of the licorice fern tastes like licorice, hence its name. It was a favorite treat of Native peoples as well as a medicinal plant.

Northern Maidenhair Fern *Adiantum pedatum*

Other Names Also known as American maidenhair fern, finger fern, hair fern, maidenhair fern, rock fern, sweet fern; formerly known as *A. aleuticum*.

Maidenhair Ferns (Pteridaceae)

Description **Overall:** Small fern that often forms colonies in suitable habitats. **Sori:** Oblong, along the edges of the upper lobe leaflets. **Leaves:** Palmate with black to red-brown stipes, top of leaf stalk twice divided; blades nearly at right angles to the leaf stalk, leaflets oblong or fan-shaped. **Other:** Plants arise from scaly rhizomes.

Size To 24" (60 cm) high.

Habitat Moist, rocky areas, in the spray zone of waterfalls and on rocks near the ocean; low to mid-elevation.

Range Alaska south to central California and east to New England.

Medicinal Uses Maidenhair fern has been used by herbalists for many years to make cough medicine. All maidenhair ferns are reputed to be excellent for treating cough, asthma and pleurisy. A leaf tea has been used in the treatment of colds, coughs and hoarseness. The rhizome has been used to soothe sore throat and to loosen phlegm. This fern is generally considered a gentle diuretic and good for jaundice and for gravel and other kidney problems. Because all maidenhair ferns are considered weak medicinally, they are best used fresh.

Precautions Be careful collecting this species, as plants are sometimes found on slippery rocks.

Notes Under the right conditions, maidenhair fern can produce beautiful and impressive stands that cover large areas.

Meadow Death-camas *Toxicoscordion venenosum*

Other Names Also known as death camas, poison camas; formerly known as *Zigadenus venenosus* (original spelling), *Zygadenus venenosus*. **Lily Family (Liliaceae)**

Description **Overall:** Perennial herb. **Flowers:** Creamy white; yellowish green oval spots (glands) at the base of the petals; clusters, foul-smelling. **Fruits:** Small, dry cylindrical capsules, form above flower. **Leaves:** Primarily basal, linear and grass-like. **Other:** Plant arises from a scaly bulb.

Size To 24" (60 cm) high.

Habitat In dry basins, open forests and forest edges, damp (in spring) meadows and grassy slopes; low to mid-elevation.

Range Central BC to Mexico.

Medicinal Uses Poultices were made for external use only, for treating bruises, boils, sprains and rheumatism.

Warnings The bulb and leaves are poisonous to humans and animals, as both its common and scientific names suggest. Symptoms of poisoning include vomiting, lowered body temperature, difficulty breathing and finally coma.

Notes This plant is so deadly that Native peoples used its bulbs to make arrow poison! Know this plant before you harvest wild onions (see p. 39) or common camas (see p. 40). The bulbs are similar-looking and a mistake can be deadly. Aboriginal peoples are known to have removed this plant from areas where they harvested food plants.

Arnica *Arnica* spp.

Heart-leaved arnica.

Aster Family (Asteraceae)

Description **Overall:** Perennial herb. **Flowers:** Yellow, sunflower-like, to 2.4" (6 cm) across; ray florets 8–16, showy; usually solitary; June to August. **Fruits:** Achenes, to 0.3" (8 mm) long, with short white hairs. **Leaves:** Always opposite, to 4" (10 cm) long, toothed margins. **Other:** Plants arise from horizontal rhizomes.

Size To 24" (60 cm) high.

Habitat Moderate to subalpine.

Range Alaska to México.

Warnings Arnica has been used medicinally but is best avoided due to its poisonous properties. If arnica enters the bloodstream, it is **toxic**. Some people react to arnica with a rash.

Notes There are many species of arnica in North America, and they all have similar properties. Two species are common in the Northwest (see below). All arnicas have bright yellow sunflower-like flowers with opposite leaves. Other yellow wildflowers such as senecios and balsamroots have alternate leaves.

Similar Species **Heart-leaved Arnica *Arnica cordifolia*** displays heart-shaped leaves, and flowers with pointed rays that are fringed at the tips.

Mountain Arnica *Arnica latifolia* displays rounded to elliptical leaves, and flowers with toothed rays that have squared-off tips.

List of Therapeutic Plant Uses
Historical use only (Results not proven)

abortion
false Solomon's seal
abscesses
single delight
aching body or muscles
cow-parsnip
lodgepole pine
allergies
Canada goldenrod
common Labrador tea
salal
anemia
common dandelion
angina
black hawthorn
anti-inflammatory agent
ocean spray
one-sided wintergreen
antibacterial
tall Oregon-grape
western redcedar
antifungal
western redcedar
antiseptic
lodgepole pine
tall Oregon-grape
appetite
chicory
false Solomon's seal
ox-eye daisy
red clover
tall Oregon-grape
arteriosclerosis
black hawthorn
asthma
balsam poplar
northern maidenhair fern

ox-eye daisy
palmate coltsfoot
round-leaved sundew
subalpine fir
western skunk cabbage
athlete's foot
red clover
western redcedar
arthritis
Canada buffaloberry
common dandelion
cow-parsnip
field mint
western bracken fern
willow
bacterial ailments
western larch
birth control
star-flowered Solomon's seal
bladder problems
common bearberry
common plantain
one-sided wintergreen
western redcedar
bleeding
common yarrow
horsetail
one-sided wintergreen
blisters
single delight
blood pressure
balsam poplar
black hawthorn
common dandelion
blood tonic
common bearberry

boils
balsam poplar
common yarrow
devil's club
fireweed
red clover
self-heal
single delight
western sword fern
bronchial inflammation
ox-eye daisy
bronchitis
balsam poplar
common yarrow
red clover
round-leaved sundew
bruises and contusions
balsam poplar
palmate coltsfoot
pearly everlasting
burns
alder
common cattail
common plantain
common snowberry
common yarrow
nodding onion
ocean spray
Pacific madrone
paper birch
pearly everlasting
salal
western sword fern
white-bark pine
willow
cancer
common dandelion
cow-parsnip
single delight
western sword fern
western yew

willow
catarrh
wild sarsaparilla
cervical erosion
one-sided wintergreen
cholagogue
chicory
cholera
fireweed
colds
common Labrador tea
common yarrow
devil's club
false Solomon's seal
field mint
licorice fern
northern maidenhair fern
Pacific madrone
pearly everlasting
pineapple weed
roseroot
single delight
Sitka mountain ash
subalpine fir
western larch
wild sarsaparilla
colitis
bunchberry
colon troubles
fireweed
conjunctivitis
ox-eye daisy
constipation
alder
cascara
chicory
common dandelion
tall Oregon-grape
Canada goldenrod
contraception
tall Oregon-grape

contraception, post-childbirth
Pacific madrone
corns and bunions
round-leaved sundew
coughs
common yarrow
devil's club
field mint
licorice fern
northern maidenhair fern
ox-eye daisy
pearly everlasting
red clover
red osier dogwood
round-leaved sundew
single delight
subalpine fir
trembling aspen
western skunk cabbage
white-bark pine
cramps
western skunk cabbage
cuts and abrasions
Canada buffaloberry
common plantain
common snowberry
lodgepole pine
one-sided wintergreen
ox-eye daisy
self-heal
subalpine fir
willow
diabetes
balsam poplar
common dandelion
devil's club
diarrhea
alder
bunchberry
Canada goldenrod
common bearberry

common cattail
common Labrador tea
common yarrow
fireweed
ocean spray
pearly everlasting
pineapple weed
red osier dogwood
salal
willow
digestive ailments
field mint
nodding onion
pearly everlasting
red clover
willow
digestive impurities
red clover
disinfectant
lodgepole pine
diuretic
northern maidenhair fern
one-sided wintergreen
dysentery
bunchberry
fireweed
pearly everlasting
earache
common yarrow
eczema
balsam poplar
fireweed
red clover
edema
ox-eye daisy
esophagus inflammations
shrubby cinquefoil
eye problems
chicory
common bearberry
common snowberry

common yarrow
ocean spray
Pacific madrone
pineapple weed
roseroot
silverweed
single delight
Sitka mountain ash
fever
broad-leaved stonecrop
bunchberry
common snowberry
common yarrow
devil's club
field mint
pineapple weed
shrubby cinquefoil
subalpine fir
trembling aspen
wild sarsaparilla
flatulence
pineapple weed
flu (influenza)
shrubby cinquefoil
single delight
western larch
fungal infections
common Labrador tea
western redcedar
gallbladder problems
ox-eye daisy
gastric ulcers
common plantain
gastritis
bunchberry
salal
gastrointestinal disorders
common plantain
general malaise
alder
chicory

ocean spray
tall Oregon-grape
gout
red clover
wild sarsaparilla
gum problems
cow-parsnip
horsetail
subalpine fir
gynecological disorders
one-sided wintergreen
hair tonic
common bearberry
wild rose
headache
balsam poplar
cow-parsnip
field mint
Sitka mountain ash
willow
heart
black hawthorn
cascara
self-heal
hemorrhage
self-heal
hemorrhoids (piles)
alder
western redcedar
hepatic
chicory
hepatitis C
western larch
HIV/aids
western larch
hives
common dandelion
inflammations
bunchberry
chicory
wild sarsaparilla

insect bites
common plantain
common yarrow
one-sided wintergreen
paper birch
salal
willow
insect infestations
lodgepole pine
internal bleeding
common plantain
shrubby cinquefoil
stinging nettle
intestinal gas
wild sarsaparilla
itchy skin
alder
blue clematis
jaundice
alder
chicory
northern maidenhair fern
trembling aspen
kidney problems
common bearberry
common plantain
field mint
northern maidenhair fern
one-sided wintergreen
tall Oregon-grape
western redcedar
kidney stones
common dandelion
leg ulcers
blue clematis
lice
common Labrador tea
Sitka mountain ash
limbs aching
cow-parsnip
liver problems

common dandelion
ox-eye daisy
red clover
western yew
lung problems
stinging nettle
menstrual problems
common snowberry
one-sided wintergreen
pineapple weed
wild rose
migraines
blue clematis
morning sickness
common snowberry
mouth problems
horsetail
ox-eye daisy
strawberry blite
mucous membranes
fireweed
nerve growth
cow-parsnip
nervous excitability
ox-eye daisy
night sweats
ox-eye daisy
nosebleed
balsam poplar
pain relief
bunchberry
field mint
lamb's quarters
salal
single delight
Sitka mountain ash
painful joints
blue clematis
paralysis
single delight

pleurisy
northern maidenhair fern
pneumonia
lodgepole pine
poison, antidote or aid
bunchberry
common plantain
one-sided wintergreen
wild sarsaparilla
poison ivy
alder
psoriasis
balsam poplar
red clover
purgative
silverweed
Sitka mountain ash
rashes
chicory
fireweed
one-sided wintergreen
red clover
willow
relaxation
pineapple weed
wild rose
respiratory ailments
alder
common plantain
licorice fern
rheumatoid arthritis
common yarrow
devil's club
field mint
lodgepole pine
Pacific madrone
palmate coltsfoot
pearly everlasting
red clover
Sitka mountain ash
tall Oregon-grape

wild sarsaparilla
ringworm
western redcedar
wild sarsaparilla
scabies
common Labrador tea
scalds
balsam poplar
nodding onion
scrofula
wild sarsaparilla
sexually transmitted disease (STD)
alder
black hawthorn
Canada buffaloberry
common snowberry
paper birch
yellow pond lily
skin problems
alder
common bearberry
common Labrador tea
fireweed
horsetail
one-sided wintergreen
ox-eye daisy
red clover
self-heal
subalpine fir
trembling aspen
white-bark pine
wild sarsaparilla
sleep
western yew
wild rose
sores
balsam poplar
blue clematis
common plantain
common snowberry
common yarrow

horsetail
nodding onion
palmate coltsfoot
pearly everlasting
subalpine fir
western sword fern
sores, weeping
Canada goldenrod
paper birch
spasms
ox-eye daisy
spleen problems
chicory
sprains
palmate coltsfoot
sterility
one-sided wintergreen
stomach ailments
common Labrador tea
common plantain
devil's club
field mint
licorice fern
Pacific madrone
palmate coltsfoot
pearly everlasting
pineapple weed
shrubby cinquefoil
stinging nettle
tall Oregon-grape
wild sarsaparilla
sunburns
balsam poplar
pearly everlasting
swellings
chicory
pearly everlasting
single delight
swollen legs
cow-parsnip
throat problems

common Labrador tea
common yarrow
fireweed
horsetail
licorice fern
northern maidenhair fern
palmate coltsfoot
pearly everlasting
roseroot
self-heal
single delight
strawberry blite
western sword fern
tonsillitis
western sword fern
toothache
common plantain
cow-parsnip
field mint
western redcedar
ulcers
fireweed
ox-eye daisy
red clover
subalpine fir
western yew
willow
urinary tract irritations
Canada goldenrod
willow
varicose veins
common yarrow
viral ailments
western larch
vocal cord swelling
ox-eye daisy
vomiting
field mint
warts
broad-leaved stonecrop
round-leaved sundew

western redcedar
worms
broad-leaved stonecrop
common plantain
lodgepole pine
trembling aspen
wounds
alder
balsam poplar
common snowberry
false Solomon's seal
horsetail
lodgepole pine
nodding onion
Pacific madrone
subalpine fir
tall Oregon-grape

Plant Glossary

achene: a small nut-like fruit with a single seed
anther: area of stamen where pollen is produced
annual: a plant that completes its entire life cycle in one year
basal: at the base
biennial: a plant that germinates in its first year and flowers, produces seeds and dies in its second year
bract: modified leaf found beneath a flower
bulb: modified underground stem with thick leaves (like an onion)
calyx: a structure made by the union of sepals
compound leaf: a leaf with two or more leaflets
coniferous: having reproductive organs in cones
corm: a bulb-like thickening of the stem used for storage
deciduous: a plant that looses its leaves annually
dioecious: male and female flowers are found on separate plants
disc flowers: the flowers in the aster family (usually tube-shaped)
follicle: fruit capsule
panicle: pyramid-shaped cluster
pappus: a modifiey calyx in the aster or composite family that includes scales, bristles or hairs
pistil: the female reproductive structures
raceme: elongated flower stalk
ray flowers: the flattened flowers in the aster family (often marginal)
rhizome: an underground stem with buds or nodes
sori: clusters of sporangia that are found on the underside of fern fronds
stigma: the tip of the pistil where pollen lands
stolon: runner with roots at the nodes
taproot: the primary root (like a carrot)
tepals: sepals or petals

Medicinal Glossary

alterative: a substance that gradually acts to nourish and improve the system

anodyne: an agent that reduces pain by reducing the sensitivity of the nerves, locally or over the whole body

anti-inflammatory: a substance that prevents or reduces inflammation

antioxidant: an agent added to a product to prevent or delay its deterioration by exposure to oxygen in air

antiscorbutic: a substance that is effective in the prevention or relief of scurvy

antiseptic: a substance that inhibits the growth of microorganisms

antispasmodic: a substance that relieves cramps or spasms of the stomach, intestines and bladder

astringent: a substance that causes contraction of tissues, arrest of secretion or control of bleeding

carminative: an agent that relieves flatulence

cathartic: a medicine that causes the bowels to be purged

cytotoxic: the degree to which an agent can bring about a specific destructive action on certain cells, or the possession of such action

decoction: an extract made by boiling plant material in water

demulcent: an oily substance that soothes inflamed mucous membranes

depurative: an agent that purifies or purges by removing waste products

diaphoretic: a substance that produces or increases sweating

diuretic: a substance that helps reduce the amount of water in the body

emetic: a substance that causes vomiting

emmenagogue: a substance that increases menstrual flow

essential oil: a natural oil obtained by distillation and having the characteristic odor of the plant or other source from which it is extracted

expectorant: a substance that loosens and clears mucous and phlegm from the respiratory tract

febrifuge (antipyretic): a substance that reduces fever

hemostatic: a substance or device that arrests bleeding

nephritic: of or related to the kidneys; renal

pectoral: helpful in relieving chest disorders
sedative: a substance having a calming or tranquilizing effect
soothe: allay, balm, becalm, calm, quiet, still or tranquilize
stimulant: a substance that produces a temporary increase of functional activity or efficiency
stomachic: beneficial to or stimulating digestion in the stomach
styptic: contracting the tissues or blood vessels, thereby stopping or slowing bleeding
sudorific: producing or increasing sweating
tonic: a substance that gives a feeling of vigor or well-being
vulnerary: used to heal a wound

Bibliography

Edible and Medicinal Plants

Angier, Bradford, and David K. Foster. 2008. *Field Guide to Medicinal Wild Plants*, 2nd ed. Mechanicsburg PA: Stackpole Books.

Hutchens, A.R. 1991. *Indian Herbalogy of North America*. Boston MA: Shambhala Publications.

Kershaw, Linda. 2000. *Edible and Medicinal Plants of the Rockies*. Edmonton AB: Lone Pine Publishing.

Krochmal, Arnold & Connie. 1973. *A Guide to the Medicinal Plants of the United States*. New York: Quadrangle/New York Times Book Co.

Marles, Robin J., et al. 2000. *Aboriginal Plant Use in Canada's Northwest Boreal Forest*. Vancouver: UBC Press.

Moore, Michael. 2011. *Medicinal Plants of the Pacific West*. Santa Fe NM: Museum of New Mexico Press.

Smith, Harlan I. 1997. *Ethnobotany of the Gitksan Indians of British Columbia*. Canadian Ethnology Service, Paper 132. Hull QC: Canadian Museum of Civilization.

Szczawinski, Adam F. and George A. Hardy, 1972. *Guide to Common Edible Plants of British Columbia*. Victoria BC: British Columbia Provincial Museum, Department of Recreation and Conservation, Handbook No. 20.

Turner, Nancy J. 2010. *Food Plants of Coastal First Peoples*. Victoria BC: Royal British Columbia Museum.

Turner, Nancy J. 2010. *Food Plants of Interior First Peoples*. Victoria BC: Royal British Columbia Museum.

Turner, Nancy J., et al. 1980. *Ethnobotany of the Okanagan-Colville Indians of British Columbia and Washington*. Victoria BC: British Columbia Provincial Museum, Occasional Papers, No. 21.

Turner, Nancy J., et al. 1990. *Thompson Ethnobotany Knowledge and Usage of Plants by the Thompson Indians of British Columbia*. Victoria BC: Royal British Columbia Museum, Memoir No. 3.

Willard, Terry. 2003. *Edible and Medicinal Plants of the Rocky Mountains and Neighbouring Territories*. Calgary AB: Wild Rose College of Natural Healing.

Plant Identification

Kershaw, Linda, A. MacKinnon and J. Pojar. 1998. *Plants of the Rocky Mountains*. Edmonton AB: Lone Pine Publishing.

Kozloff, Eugene. 2005. *Plants of Western Oregon, Washington, and British Columbia*. Portland OR: Timber Press.

Lyons, C.P. 2000. *Wildflowers of Washington*. Edmonton AB: Lone Pine Publishing.

Lyons, C.P., and B. Merilees. 1996. *Trees, Shrubs & Flowers to Know in Washington & British Columbia*. Edmonton AB: Lone Pine Publishing.

MacKinnon, Andy, J. Pojar and R. Coupe. 2005. *Plants of Northern British Columbia*, 2nd ed. Edmonton AB: Lone Pine Publishing.

Parish, Roberta, et al. 1999. *Plants of Southern Interior British Columbia and the Inland Northwest*, 2nd ed. Edmonton AB: Lone Pine Publishing.

Pojar, Jim, and A. MacKinnon. 2004. *Plants of Coastal British Columbia Including Washington, Oregon and Alaska*, rev. ed. Edmonton AB: Lone Pine Publishing.

Sept, J. Duane. 2005. *Wild Berries of the Northwest: Alaska, Western Canada & the Northwestern United States*. Sechelt BC: Calypso Publishing.

Sept, J. Duane. 2011. *Trees of the Northwest: Alaska, Western Canada & the Northwestern United States*. Sechelt BC: Calypso Publishing.

Acknowledgements & Credits

I would like to thank a few people who assisted with this project.
Mary Schendlinger for her careful and insightful editing.
Jim Salt, who generously aided me in locating several species for photography.

All photographs by J. Duane Sept except for the photograph of the author by Susan Servos-Sept on p. 95.

Index

About the Author

Duane Sept is a biologist, freelance writer and professional photographer. His biological work has included research on various wildlife species and service as a park naturalist. His award-winning photographs have been published internationally, in displays and in books, magazines and other publications, for clients that include BBC Wildlife, Parks Canada, Nature Canada, National Wildlife Federation and World Wildlife Fund.

Today Duane brings a wealth of information to the public as an author, in much the same way he has inspired thousands of visitors to Canada's parks. His published books include *The Beachcomber's Guide to Seashore Life in the Pacific Northwest* (Harbour Publishing), *Trees of the Northwest: Alaska, Western Canada and the Northwestern United States* (Calypso Publishing), *Wild Berries of the Northwest: Alaska, Western Canada and the Northwestern United States* (Calypso Publishing) and *Common Mushrooms of the Northwest: Alaska, Western Canada and the Northwestern United States* (Calypso Publishing).

More Great Nature Books from
Calypso Publishing

Wild Berries of the Northwest:
Alaska, Western Canada and the Northwestern United States
J. Duane Sept

Fruits and berries are all around us. Identify these fruits and their flowers on your next trip to the ocean, lake or woods with this full-color guide. Learn which species are edible and which are poisonous. An entire chapter of mouth-watering recipes is also featured. Enjoy!

5.5" x 8.5" • 96 pages • 169 color photos
Softcover • $14.95 • ISBN 978-0-9739819-3-3

Common Wildflowers of British Columbia
J. Duane Sept

This easy-to-use, beautifully illustrated guide helps you identify 142 species of wildflowers that grow in British Columbia. Lavish full-color photos and clear, understandable descriptions illuminate each of the common and uncommon flowers. Also included are little-known facts, edible species, notes on aboriginal use and a checklist.

5.5" x 8.5" • 96 pages • 145 color photos
Softcover • $12.95 • ISBN 0-9730390-9-4

Trees of the Northwest:
Alaska, Western Canada and the Northwestern United States
J. Duane Sept

Trees are all around us! Some live more than 1,500 years, others produce spectacular color displays in the autumn, still others have medicinal properties. More than 49 amazing species, accompanied by more than 190 full-color photos, are featured in this concise, attractive guide. Now it is easier than ever to identify—and appreciate—our fascinating trees.

5.5" x 8.5" • 96 pages • 200+ color photos
Softcover • $14.95 • ISBN 978-0-9739819-4-0

••

These titles are available at your local bookstore or

Calypso Publishing

www.calypso-publishing.com